PRACTICAL BIOLOGY

SECOND EDITION

C. DODDS B.Sc., M.I.Biol.
Senior Biology Master
King Edward's School, Birmingham

J. B. HURN A.T.D., F.R.S.A., R.B.S.A.
Director of Art and Design
King Edward's School, Birmingham

 EDWARD ARNOLD

Preface

The aim of this book is to provide a sound basis for undertaking practical biology in the laboratory. The material covers that required by G.C.E. 'O' level syllabuses, but at the same time is not bound completely by them.

While there are many text-books available for the theoretical aspect of the subject, the authors feel that there is a need for a good practical foundation to the study of biological principles. In large classes and with the shortage of laboratory facilities, teachers often find it difficult to encourage good practical work. Pupils cannot receive as much individual attention as is necessary for a satisfactory explanation of specimens and experiments under their observation. This book is an attempt to assist with this problem in practical classes.

We have aimed at developing the powers of observation, using the three fundamental techniques of biology; dissection, microscopy and simple experiments with living organisms. In the Introduction, the methods of drawing and recording using these three techniques are given. Sections II and III cover the systematic study of animal and plant types. To help the pupil to a full understanding of structure, we have been concerned to indicate size and the form of various organisms as seen in three dimensions. It is not intended that drawings and diagrams should be copied into note-books, but rather that they should serve as a guide to material under observation and so help pupils to make good records themselves. A note is given on most species indicating the methods which should be employed to prepare them for observation in class.

Section IV covers a wide range of experiments involving observations of the main principles of biology, those described have been successfully used with the 11–16-year-old age range.

We wish to express our appreciation for the help and advice we have received from many friends, and in particular H. W. Ballance, M.A., L. H. Finlayson, Ph.D., B.Sc. and G. L. Parker. Our thanks are also due to Messrs. A. R. Leech, A. J. Markham and S. Minton for assistance in preparing specimens and labelling diagrams and drawings. We thank the following for their kindness in granting permission to base diagrams on their apparatus: Vickers Instruments Ltd. and W. Watson and Sons Ltd. for the microscopes in Section I; T. Gerrard and Co. Ltd. and Mr. D. Etherington for the osmometer in Section 20.2.2. Finally, we are grateful to our publishers for their help in the preparation of this book.

1964

C.D.
J.B.H.

Preface to the Second Edition

Although it is often said that it is difficult to combine the various aspects of modern biology into one book even at elementary level, we are encouraged by the response to the first edition to believe that there is a need for a compact practical text-book. We have taken the opportunity in this Second Edition to increase the range of organisms in the Zoology and Botany Sections. The Practical Section has been considerably expanded to include up-to-date techniques. All the sections have been set out in a logical order, and the system of cross-referencing quickly relates physiological experiments in Section IV to Sections II and III where the particular organisms used as experimental material are described. In revising the text, S.I. Units have been incorporated.

Some of the experiments in Section IV have been influenced by the work of the Nuffield Foundation, and in particular we wish to acknowledge their permission to use the following experiments:

Vitamin C tests—page 104
Visking Tubing Osmometer—page 106
Water loss from leaves—page 108
Gas analysis apparatus—page 109
Tullgren funnel—page 111
Locust testis preparation—page 118.

1971

C.D.
J.B.H.

© C. Dodds and J. B. Hurn, 1972

First published 1964
by Edward Arnold (Publishers) Ltd.,
41 Maddox Street,
London W1R 0AN

Reprinted 1965
Reprinted 1968
Second Edition 1972

ISBN: 0 7131 1699 4

Printed by offset in Great Britain by
William Clowes & Sons, Limited, London, Beccles and Colchester

Contents

I INTRODUCTION

1.1 Practical biology

Biology is a science which studies the structure of living things, and the manner in which they function. In order to do this, it is necessary to observe a wide variety of organisms, and to make a study of them in a systematic manner. These observations can often be carried out in natural surroundings out of doors, but it is more convenient to work through a great deal of such study indoors in a laboratory.

Practical work requires the ability and patience to observe carefully, and to record the results of these observations by making brief notes, drawings or diagrams.

No two organisms are exactly alike, so you should always make your own observations carefully, and make your own records. The notes and diagrams in this book are intended to help you with your practical work, but you should not attempt to copy these. Instead you should use them to help you to find out more about the various organisms which are described.

There are three main kinds of practical work in biology.

(a) Microscopic work, to observe the details of the structure of plants and animals, which can only be observed clearly when magnified by a microscope.

(b) Dissection, a method of carefully separating a plant or animal into its main parts, to see how they fit together.

(c) Experiments, in which living organisms are studied in order to observe how they behave either in their natural surroundings or in a laboratory under conditions which can be controlled by the observer.

1.2 Method of studying organisms

Living organisms have a variety of structures which enable them to carry out their life processes. It is best to study these in a systematic manner, and throughout the book you will find brief notes under the headings of the main living processes: Irritability, Movement, Nutrition, Respiration, Excretion and Reproduction.

1.3 Classification

There are many million different types of animals and plants, and they are arranged into various groups and sections in the form of a systematic catalogue. The main subdivisions are:

Kingdom Animals, or Plants.

Phylum A large group of Animals, the individuals in the group being very similar in their general structure. E.g. *Arthropoda*, animals with jointed limbs, and an external skeleton.

Class A major subdivision of a Phylum. E.g. *Insecta*, jointed-limbed animals with 6 legs.

Order A major subdivision of a Class. E.g. *Diptera*, 2-winged insects, such as flies.

Family A major subdivision of an Order. E.g. *Muscidae*, house flies, a group of organisms very similar in structure and appearance.

Genus A subdivision of a Family, and refers to a very definite type of organism. E.g. *Musca*, the house fly.

Species The main unit in the classification scheme. Organisms of the same species will breed together naturally, and reproduce their own kind. E.g. *Musca domestica*, the common house-fly. An organism is named with a form of Latin, the genus name first (with a capital letter) followed by the species name (with a small letter). This system of double names is therefore rather like the method of surnames and Christian names used in naming human beings.

This system of naming and classification is used throughout this book. In some cases, the notes describing some of the plants and animals apply to the various species within a particular genus. Where this occurs, the letters spp. (the abbreviated plural of the word species), appear after the generic name. E.g. on page 15, *Planaria* spp.

4

2 THE MICROSCOPE

This is one of the most important and most expensive instruments in the biology laboratory, and its careful and correct use makes it possible to observe the inner structure of organisms, magnified to approximately 1,000 times. Always keep it clean, and put it away carefully after use.

The standard microscope

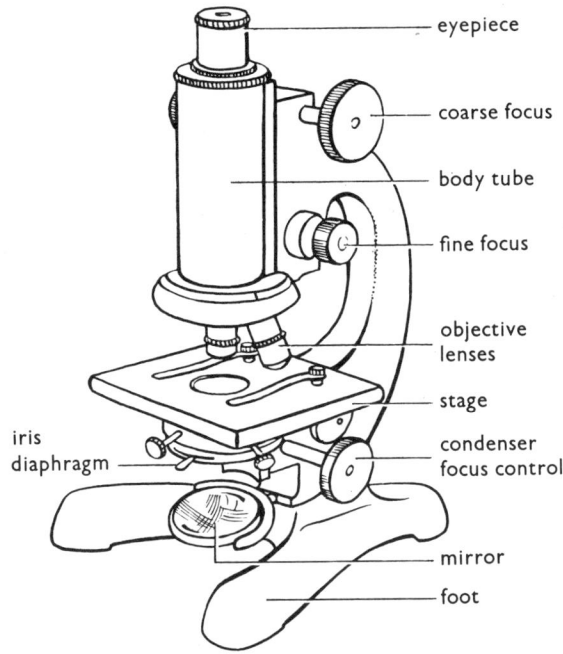

Student microscope with built-in illumination

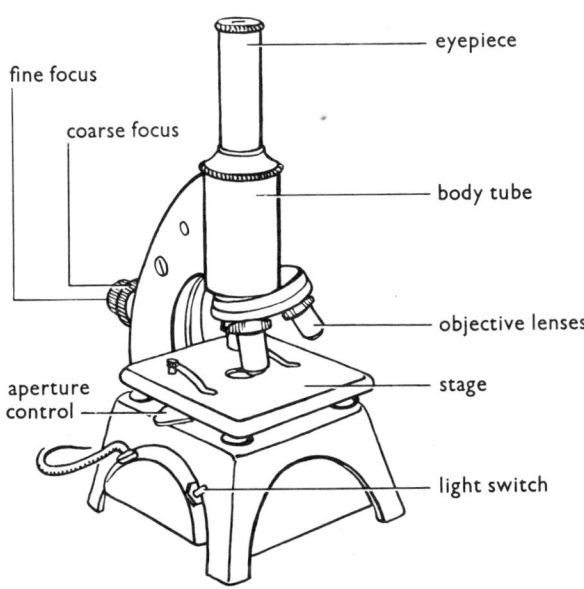

2.1 The standard microscope

Identify the main parts of the instrument, and learn how to use it.

2.2 Rules for its use

(a) See that the stage is horizontal, and the low power objective lens in position.

(b) Look through the eyepiece with one eye; try to keep the other one open at the same time. Adjust the mirror so that light reaches the eyepiece, either from a window or bench lamp. Adjust the iris diaphragm so that the light reaching your eye is not too bright.

(c) Place the object to be examined on a microscope slide. The object must be transparent. (See 2.6.)

(d) Put a drop of water on the object, and cover it carefully with a coverslip. (See 2.6.)

(e) Looking at the side of the microscope, adjust the low power lens so that it is about 1 cm away from the object. Then looking through the eyepiece again, focus the body of the microscope upwards with the coarse adjustment wheel, until the object comes into focus.

(f) Control the light with the iris diaphragm with one hand, and with the other focus with the fine adjustment wheel, until the clearest view of the object is obtained.

(g) Look to the side of the microscope again, and carefully swing the high power objective lens into position, taking care to see that it does not touch the coverslip. Then look through the eyepiece, and focus upwards gently with the fine adjustment wheel, again balancing the amount of light with the iris diaphragm.

(i) After use, put the microscope away with objective lenses in the low power position.

(j) Never touch the lenses with your fingers.

2.3 Student microscope with built-in illumination

Identify the main parts of the instrument, and learn how to use it.

2.4 Rules for its use

(a) Place the object to be examined on a microscope slide. The object must be transparent. (See 2.6.)

(b) Put a drop of water on the object, and cover it carefully with a coverslip.

(c) Place the slide on the stage, with the object as near the centre as possible, and switch on the light.

(d) Look at the side of the microscope, and adjust the low power lens, so that it is about 0·5 cm away from the object. Then looking through the eyepice, focus upwards with the coarse adjustment wheel until the object comes into focus.
Adjust the aperture control to give the right amount of light.

(e) After studying the object at a low power of magnification, look at the side of the microscope again and cautiously swing the high power objective lens into position, taking care to see that it does not touch the coverslip. Then look through the eyepiece and focus carefully with the fine adjustment wheel until the object can be seen clearly.

(f) After use, put the microscope away with the lenses in the low power position.

(g) Never touch the lenses with your fingers.

2.5 Use of hand lenses

Some larger objects are best examined with a hand lens. Place the object in a watch-glass on top of a suitable background. Usually a piece of black paper is a good background for light coloured objects. Hold the lens steady in one hand, and bring the object into focus, use the other hand to record your observations.

2.6 How to mount fresh objects on a slide for observation

It is necessary to support small fresh animals or plants, or parts of them in water, so that they can be seen under the microscope or hand lens. Pick the object up on the tip of a small paint brush, or in a pair of forceps, and place it carefully in the centre of a microscope slide. Place one drop of water on it from a small dropper or pipette, so that it is completely covered. Next, pick up a coverslip carefully with the forceps, and place it in a slanting position over the water drop, just touching the slide at its lower end. Support the upper end using a seeker or the end of a wooden spill, and lower it slowly, taking care to avoid trapping any air bubbles. Remove any surplus water extending beyond the coverslip by sucking it up with a small piece of blotting paper. Remember that when the object is viewed under the microscope, it will appear upside down, and reversed from left to right.

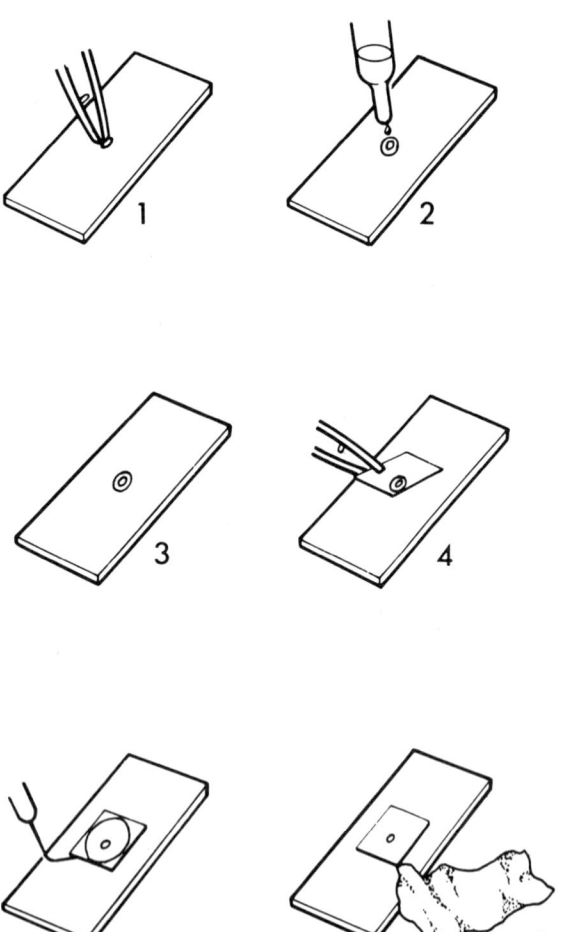

2.7 How to make a smear preparation

Place a very small drop of blood, or earthworm sperms (see section 9.1, page 19), on the centre of a slide. Touch the drop with another slide held at a slanting angle, and move it away rapidly. This action will draw a thin film along behind it. Place a drop of 3% salt solution over the film, cover with a coverslip and examine it under low and high power lenses.

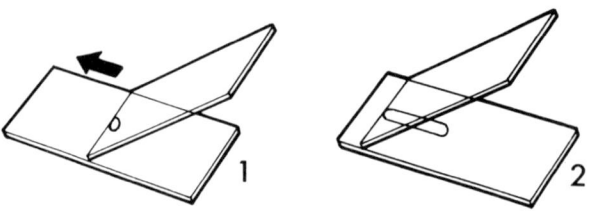

2.8 Examining prepared microscope slides

Some specimens are difficult to obtain, others are difficult to see when mounted in a fresh condition in water. For this reason microscopic slides are supplied in a more permanent form. The specimen, usually stained with dyes to make parts of it more easily visible, is fixed to the slide and covered with a coverslip which is cemented to it with a clear adhesive called Canada Balsam. Such slides should be examined under the microscope as described in sections 2.2 or 2.4. Before mounting a prepared slide on the microscope stage, make sure that it is quite clean, and if necessary remove any dust from it by wiping it gently with a duster.

2.9 Measuring the size of objects under the microscope

It is very important to realize the size of an object when seen under the microscope. A good method is to work out the diameter of the circular view observed when looking through the microscope, and then judge the size of the object being observed as compared with the diameter of the field of view. To do this, place the edge of a transparent ruler across the circular gap on the stage and view it under low power and see how many millimeter marks can be seen along the diameter. To measure the diameter of the high power view make two marks 1 mm apart with a fine pen, using Indian ink on a piece of tracing paper. Mount this dry under a coverslip, and place the slide on the stage. Focus carefully with the high power lens, and estimate the diameter of the field of view.

In giving the size of objects seen under the microscope a unit of 1/1,000 mm is used, represented by the Greek symbol μ. This unit is used in indicating the size of some objects shown in this book. (In S.I. Units this is a micrometre, symbol μm).

2.10 Examining sections of organisms

It is not always possible to examine whole objects under the microscope, especially when they are large or not transparent. In such cases the specimen is frequently cut into sections, which are transparent enough to be seen in the light which passes through them from the mirror to the eyepiece of a microscope. A transverse section (T.S.) is one cut across a structure, and is often circular in outline. A longitudinal section (L.S.) is cut along the length of a structure and is usually rectangular in appearance.

2.11 Recording observations made under the microscope

To record what is seen under the microscope it is best to make a simple diagram or drawing. Do not attempt to draw every detail you see, especially when the object contains many similar cells. First, examine the object under the low power and draw a diagram. A diagram consists of an outline of the shape of the object, and a record of its main parts. (See, for example, section 17.2.1 for diagrams of transverse sections of stems and roots.) Wherever possible diagrams and drawings should be made on plain paper.

Always put a heading above your work. Use a good pencil and keep its point sharp. Make large, clear diagrams with firm, clean lines. Do not shade. Use coloured crayons or pencils only when necessary to distinguish particular features on your diagram. A good guide to the size of your diagram is to make 1 mm length as seen under the low power of the microscope, about 10 cm long on your page.

When labelling your diagram observe the following rules:
(a) Guide lines should be drawn in pencil with a ruler.
(b) Do not use arrow heads.
(c) Where there are a few guide lines keep them parallel as far as possible. When there are many you can draw them at an angle, but not crossing each other.
(d) Do not put labelling on the diagram itself, keep your labelling at least 2 cm away from it.
(e) Label in pencil, using printed letters.

When examining objects under the high power, select a small part of the slide, and try to make an accurate drawing of that part. A drawing differs from a diagram in that it is an attempt to show an item as accurately as possible, whereas a diagram tries to give an outline of the whole object. You can indicate on your low power diagram, e.g. by a letter, the part you attempt to draw accurately under the high power. For example, when drawing cells under the high power only draw a few from any one part of the slide. A typical cell should occupy a square whose sides are 2 cm long, when recorded on your drawing page. Label your drawing, using the same rules as used for the diagram.

3 DISSECTION

3.1 Instruments required

Scissors	4 in with fine points.
Scalpel	1 in blade.
Forceps	Fine points, with guide pin to close the points together accurately.
Mounted needles	1 pair.
Seeker	
Mounted razor blade	
Pipette (dropper)	

Always treat dissecting instruments with care, and only use them for the purpose intended. When putting them away, it it is a good idea to give them a thin covering of vaseline, after they have been thoroughly cleaned.

3.2 General rules for animal dissections

(a) *Small animals less than 15 cm long.* The animal, which should be preserved or freshly killed, is fixed to the wax of a dissecting dish by means of pins. Always identify the dorsal and ventral surfaces before fixing the animals to the wax. In general, Invertebrate animals should be dissected with the ventral surface pinned to the wax. Vertebrate animals should normally be dissected with the dorsal surface pinned to the wax.

Start your dissection by lifting a small piece of skin with the forceps, and snip an opening with the points of the scissors.

Continue your dissection by cutting along a line, always pulling the points of the scissors upwards towards you, to lift the skin up. Never dig your scissors down into an animal as this is likely to damage internal organs. Pin the skin back carefully as the dissection proceeds. Details are given under the heading 'Method of Examination' for each animal described in this book, as the precise method varies for different animals.

(b) *Large animals.* In the case of larger Vertebrate animals, dissections should be made by fixing the dorsal surface of the animal to a dissecting board. Place an awl or nail firmly through each limb, and dissect using the same general rules as for smaller animals.

3.3 General rules for plant dissections

In general, the parts of plants should be carefully removed with forceps. For example, when studying the parts of a flower, remove the various structures in order, and lay them out carefully on a tile, or piece of cardboard. A pair of mounted needles can be used for teasing or tearing apart smaller structures, which can then be placed on a small tile, or microscope slide.

3.4 How to record dissections

As with the recording of microscope observations, the problem is to show as clearly as possible the structures which can be observed when the dissection is completed. In general a dissection should be recorded by a drawing which should normally occupy a whole page. With a ruler make very light construction lines indicating the main central axis of the animal and the position of any important features such as appendages. Measure their position, so as to get the proportions accurate on your drawing. Do not attempt to draw everything you can see, rather indicate the outline of the animal, and then with firm, clear lines indicate the position and shape of the main organs. The same rules for labelling should be followed, as those given for microscopic diagrams. (See 2.11.)

4 EXPERIMENTS ON ANIMAL AND PLANT BEHAVIOUR

Many useful and interesting experiments can be performed by observing the behaviour of organisms when they are alive and in a healthy condition.

4.1 Experiments

Living organisms react in a variety of ways to changes which take place in their surroundings. The way in which they behave can be studied by carrying out biological experiments in the laboratory. It is usual to keep one organism in normal conditions, and this is called the control experiment, whilst another similar organism is placed in an experiment in which one condition, for example temperature, is very different from that of the control experiment.

4.2 Observations

The results of experiments should be recorded as soon as they have been made. As a general rule, the apparatus used for both the experiment and its control should be drawn. The experiment should be described briefly, together with any conclusions reached. No two organisms behave in exactly the same way, so observe carefully how the particular organism you are studying behaves under normal and experimental conditions.

II ZOOLOGY

The Animal Kingdom consists of two main groups.

(a) **The Protozoa,** microscopic animals, often single celled.

(b) **The Metazoa**

   (i) Invertebrate animals without backbones.

   (ii) Vertebrate animals with backbones.

  Metazoans are composed of many cells, organised in layers.

The *ectoderm*, the outermost layer forms the skin tissues.

The *endoderm*, or innermost layer forms the digestive system.

DIPLOBLASTIC animals have these two layers only.

The *mesoderm* is a third layer found in more complicated Metazoans. It lies between the other two, forming such structures as muscles and blood systems.

TRIPLOBLASTIC animals have all three layers. In some Phyla the mesoderm itself exists in two layers surrounding a body cavity called the *coelom*. Triploblastic animals without a coelom are said to be acoelomate.

Chart showing the main Invertebrate Phyla

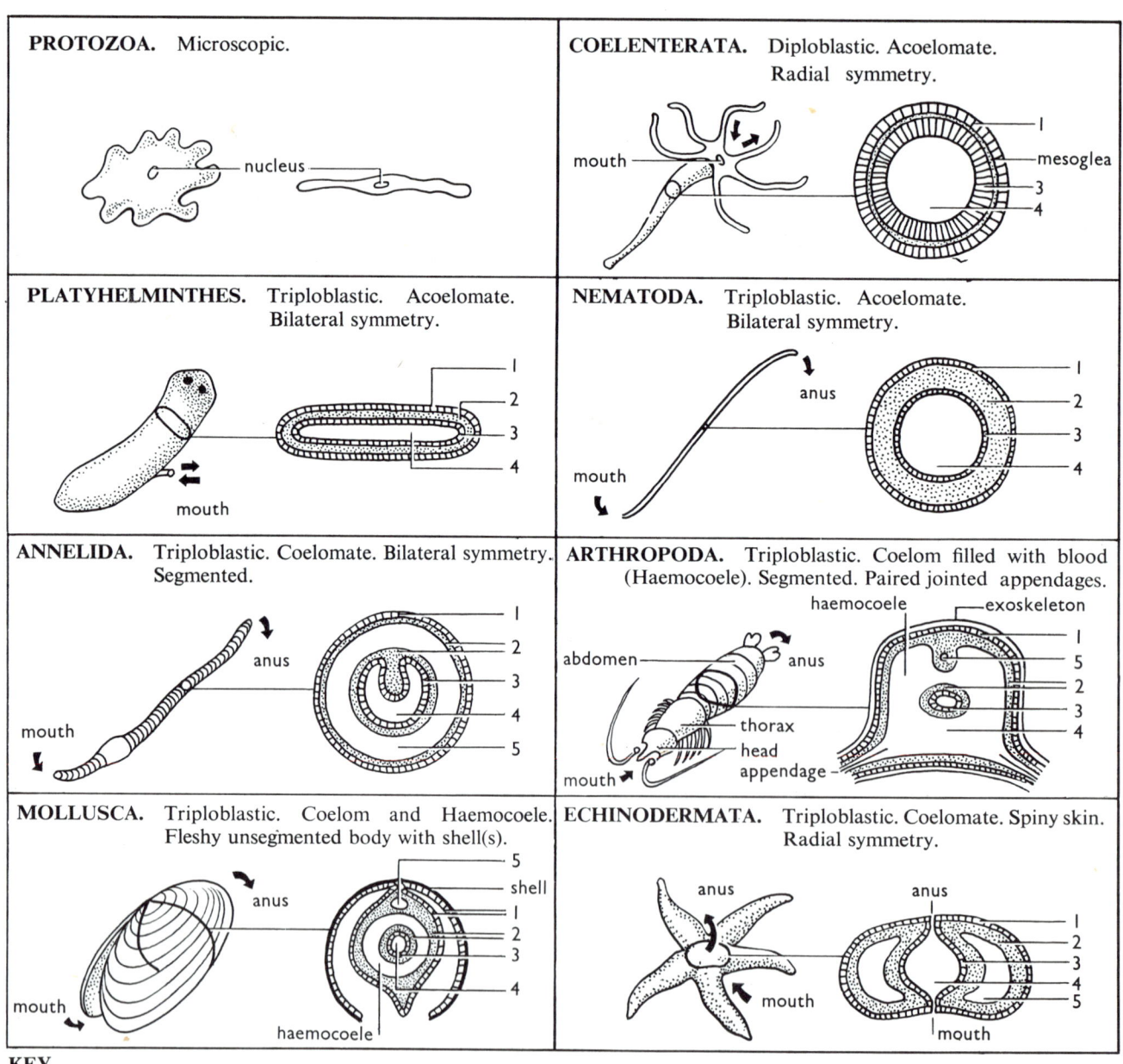

PROTOZOA. Microscopic.

COELENTERATA. Diploblastic. Acoelomate. Radial symmetry.

PLATYHELMINTHES. Triploblastic. Acoelomate. Bilateral symmetry.

NEMATODA. Triploblastic. Acoelomate. Bilateral symmetry.

ANNELIDA. Triploblastic. Coelomate. Bilateral symmetry. Segmented.

ARTHROPODA. Triploblastic. Coelom filled with blood (Haemocoele). Segmented. Paired jointed appendages.

MOLLUSCA. Triploblastic. Coelom and Haemocoele. Fleshy unsegmented body with shell(s).

ECHINODERMATA. Triploblastic. Coelomate. Spiny skin. Radial symmetry.

KEY.

1. Ectoderm. 2. Mesoderm. 3. Endoderm. 4. Enteron. (The internal cavity of the digestive system.) 5. Coelom.

5.1 *Amoeba* spp.

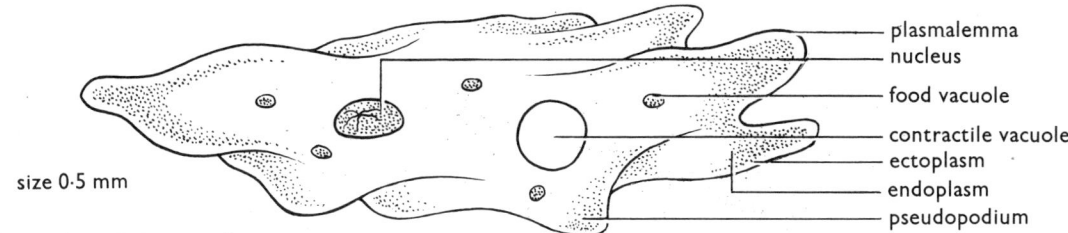

plasmalemma
nucleus
food vacuole
contractile vacuole
ectoplasm
endoplasm
pseudopodium

size 0·5 mm

Habitat Ponds, pools and damp soil.

Movement A series of drawings of an *Amoeba* taken at 1 min. intervals to show pseudopodial movement.

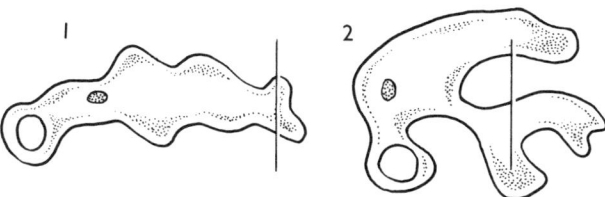

direction of movement →

Nutrition The animal feeds on microscopic organisms. An *Amoeba* devours a small organism.

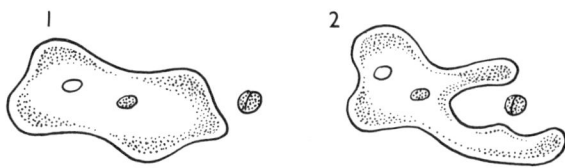

Reproduction When well fed, an *Amoeba* reproduces by dividing directly into two, the whole process being complete in an hour. Some species are also capable of reproducing by means of spores made in the endoplasm.

Encystment In unfavourable conditions, such as when a pond dries up, *Amoeba* becomes spherical in shape, and makes a cyst, in which it survives until conditions become favourable again. Within the cyst the animal may divide to produce several smaller *Amoebae*.

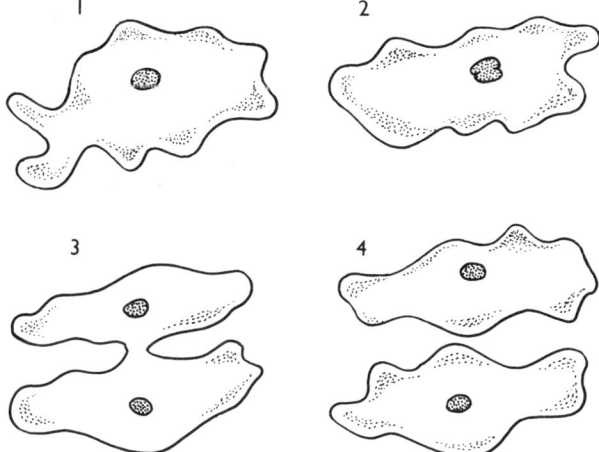

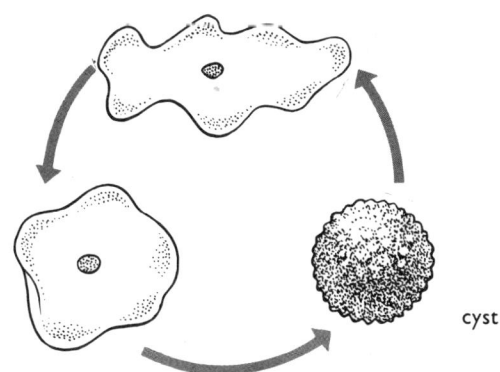

cyst

Method of examination (using the microscope) Place a drop of culture on a clean slide, and leave for 2 min. This gives the *Amoebae* time to stick to the slide. Then carefully lower a coverslip over the culture. Search for *Amoeba* using the low power lens, and using the least amount of light. *Amoeba* will be found as a grey mass, with a granular appearance. Having found a specimen, continue your examination using the high power lens. Do not let your culture dry out; add a further drop of water from a pipette when necessary.

5.2 *Paramecium* spp.

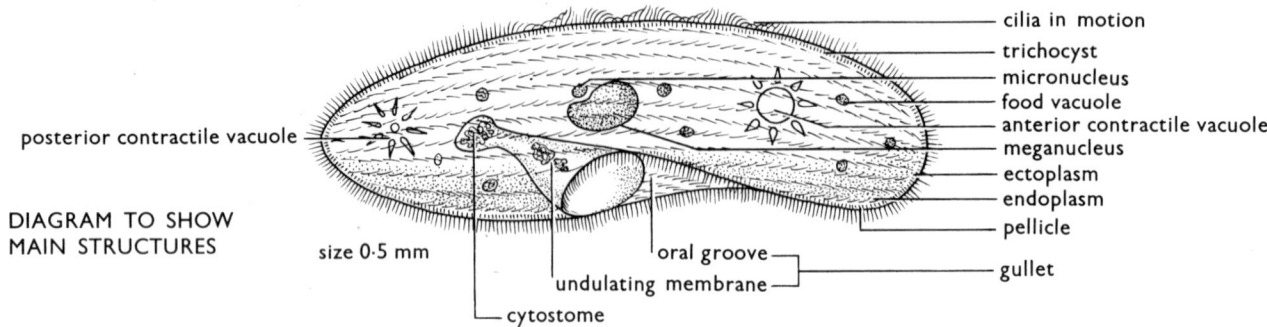

DIAGRAM TO SHOW
MAIN STRUCTURES

size 0·5 mm

Labels: posterior contractile vacuole, cilia in motion, trichocyst, micronucleus, food vacuole, anterior contractile vacuole, meganucleus, ectoplasm, endoplasm, pellicle, gullet, oral groove, undulating membrane, cytostome

Habitat Ponds, pools and damp soil.

Movement The animal spirals as it moves forward by means of the rhythmic beat of its *cilia*. This can be observed by watching the position of its *gullet*.

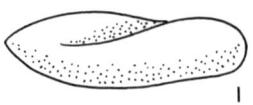
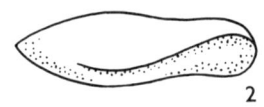

direction of movement

position of gullet

When a *Paramecium* meets an obstacle, it alternately reverses and goes forward at slightly different angles, until the obstruction is finally avoided.

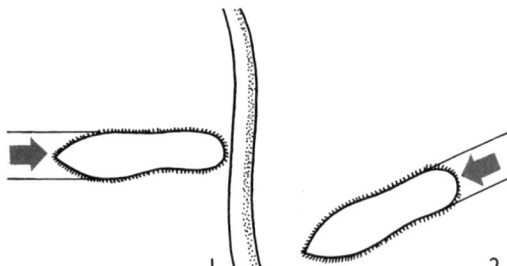
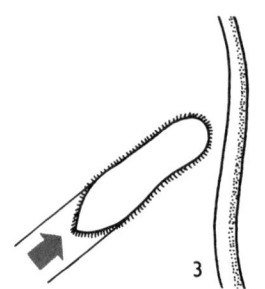
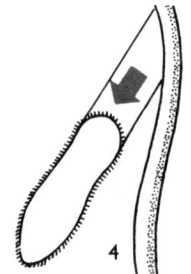
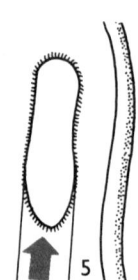

If *Paramecium* is placed in an irritant, such as very dilute acetic acid, it discharges its *trichocysts*.

Contractile vacuoles These eliminate water rhythmically, and can be watched in an animal when trapped at rest.

Stages drawn at 30 second intervals.

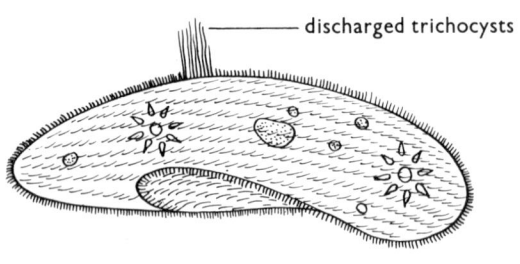

discharged trichocysts

Nutrition *Paramecium* feeds on bacteria, which it engulfs in its gullet.

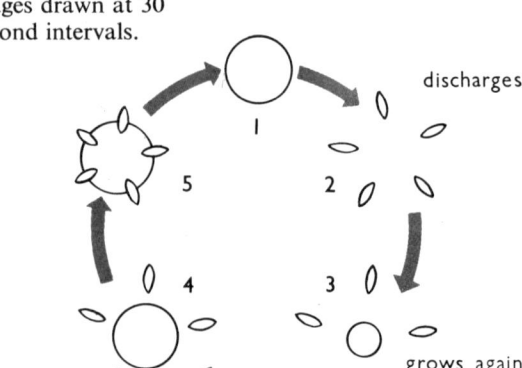

discharges

grows again

10

Reproduction

(a) Asexual The animal splits transversely, the *meganucleus* and *micronucleus* divide in two, and separate into each new half.

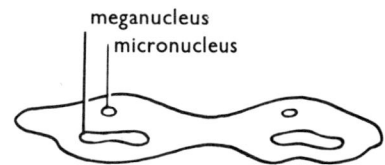

(b) Sexual (Conjugation) Two individuals join together lengthwise, their gullets being adjacent to each other. While the *meganucleus* in each individual slowly disintegrates, the *micronucleus* divides to produce four daughter nuclei.

Of these four, three degenerate and the remaining one divides again, to leave two nuclei in each individual. As shown in the diagram, one of these nuclei from each individual crosses over to fuse with the stationary nucleus in the opposite individual.

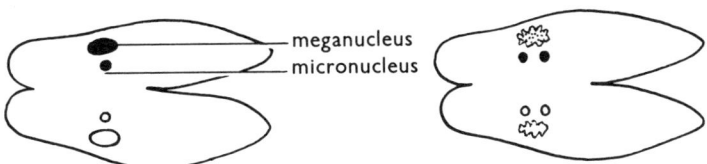

1. Two individuals join by their oral grooves.

2. The *meganucleus* degenerates. The *micronucleus* divides.

3. The *micronucleus* divides again.

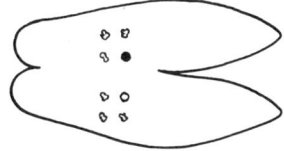 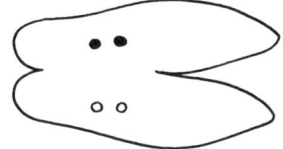 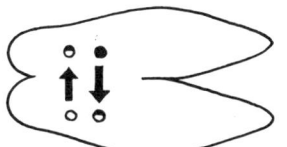

4. Of the four *micronuclei* now present in each individual, three degenerate.

5. The remaining one now divides again.

6. One nucleus from each individual fuses with one from the opposite individual.

7. The individuals then separate, and each divides to produce four small *Paramecia*. As the original nucleus divides to produce eight nuclei, four of which enlarge, each resultant *Paramecium* therefore contains two nuclei, one *meganucleus*, and one *micronucleus*.

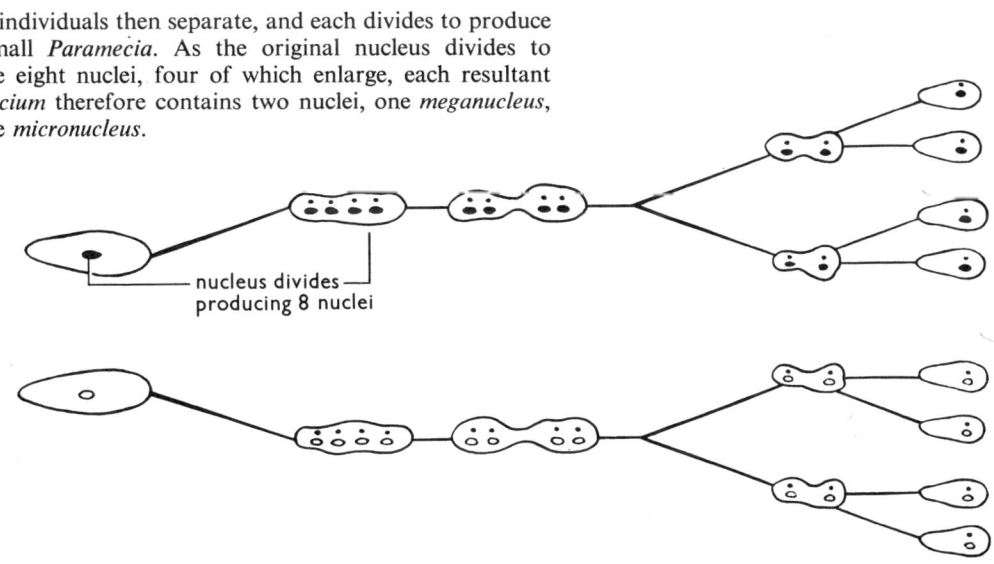

nucleus divides producing 8 nuclei

Method of examination (using the microscope) Place a drop of culture on a slide, and examine under the low power lens. To slow the animals down, place a drop of thin agar jelly on the slide, and gently cover with a coverslip.

Observe the movements of the *contractile vacuole* when the animal is at rest, using the high power lens if possible. Be careful to use the smallest amount of light. Examine prepared slides of conjugation.

6 Coelenterata

Simple sack-like animals, the opening surrounded by tentacles. The body made from two layers of cells (diploblastic), separated by a gelatinous skeleton of mesoglea.

6.1 *Hydra* spp.

EXPANDED. length 12 mm   CONTRACTED

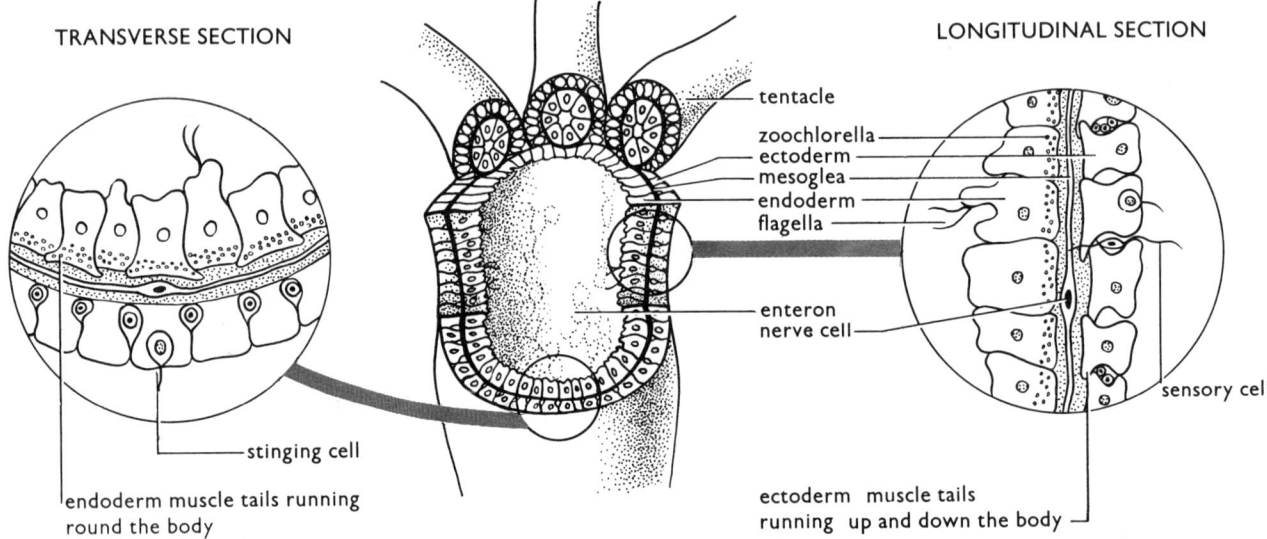

HYDRA IN THE REGION OF
THE MOUTH
EXPOSED TO SHOW THE
DIPLOBLASTIC STRUCTURE

TRANSVERSE SECTION   LONGITUDINAL SECTION

Habitat   Fresh water, usually attached to water plants.

Movement   The animal can glide on its foot, or loop. Very occasionally it is able to somersault.

Irritability   A network of nerves runs in the *mesoglea*. They are connected to sensory cells which occur frequently in the *ectoderm*, but are less common in the *endoderm*.

Nutrition   It catches its prey, small water crustacea, by numbing them with its stinging cells, and then using its tentacles to force the food into its mouth.

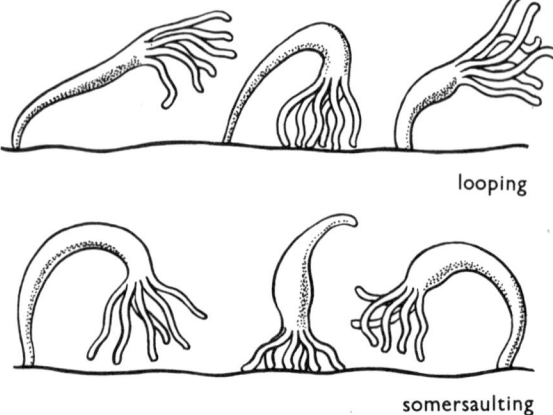

looping

somersaulting

Reproduction

(a) Vegetative In favourable conditions, buds develop and then break away from the parent to begin a separate existence.

(b) Sexual When the water in which it lives becomes stagnant, *Hydra* reproduces sexually. The animal is hermaphrodite but the sex organs on one animal ripen at different times, so that cross-fertilization between two animals is usual.

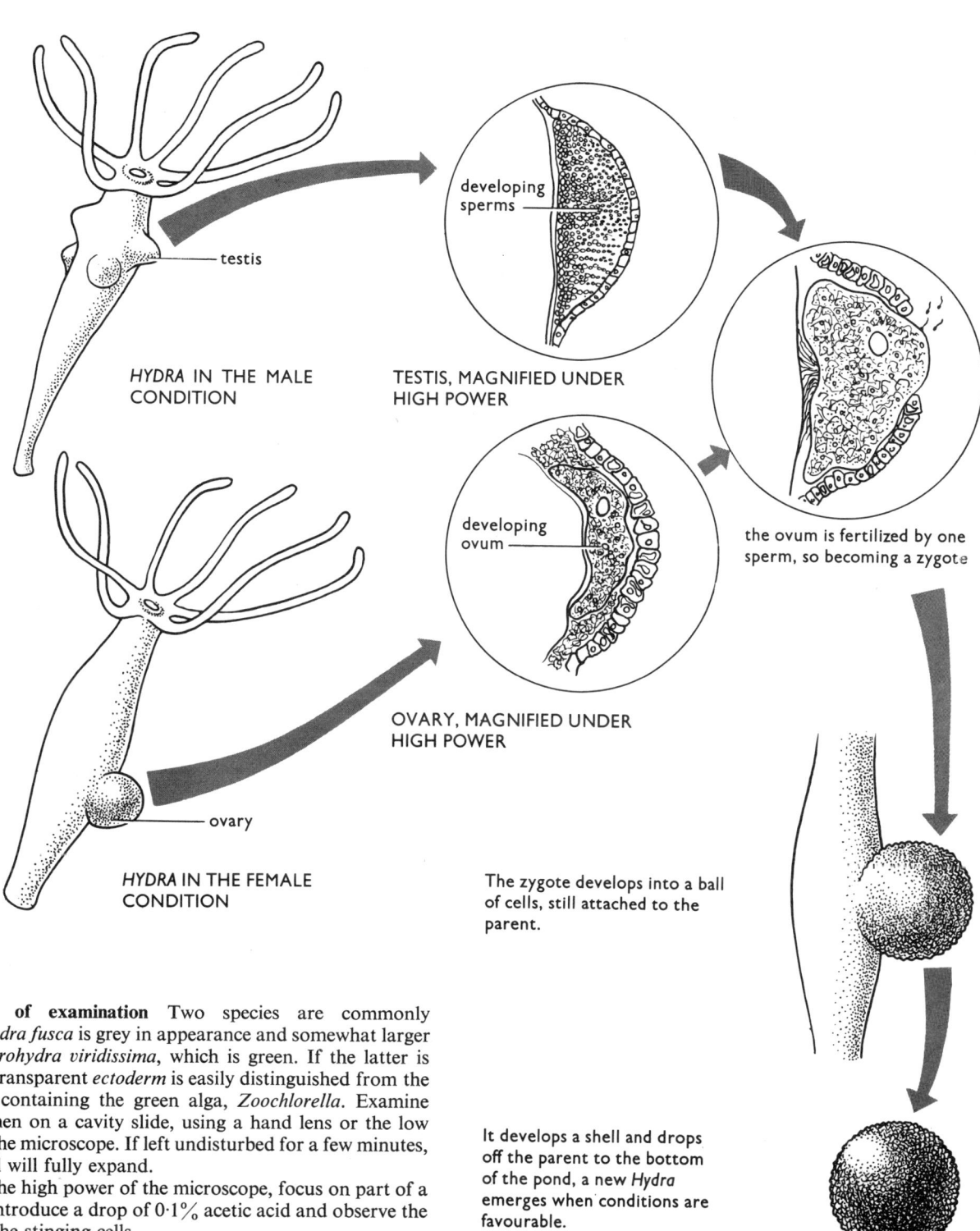

HYDRA IN THE MALE CONDITION

— testis

TESTIS, MAGNIFIED UNDER HIGH POWER

developing sperms

developing ovum

OVARY, MAGNIFIED UNDER HIGH POWER

HYDRA IN THE FEMALE CONDITION

— ovary

the ovum is fertilized by one sperm, so becoming a zygote

The zygote develops into a ball of cells, still attached to the parent.

It develops a shell and drops off the parent to the bottom of the pond, a new *Hydra* emerges when conditions are favourable.

Method of examination Two species are commonly found. *Hydra fusca* is grey in appearance and somewhat larger than *Chlorohydra viridissima*, which is green. If the latter is used, the transparent *ectoderm* is easily distinguished from the *endoderm* containing the green alga, *Zoochlorella*. Examine the specimen on a cavity slide, using a hand lens or the low power of the microscope. If left undisturbed for a few minutes, the animal will fully expand.

Under the high power of the microscope, focus on part of a tentacle; introduce a drop of 0·1% acetic acid and observe the action of the stinging cells.

Examine prepared slides of transverse sections showing testes and ovaries.

13

6.2 The Beadlet Anemone (*Actinia equina*)

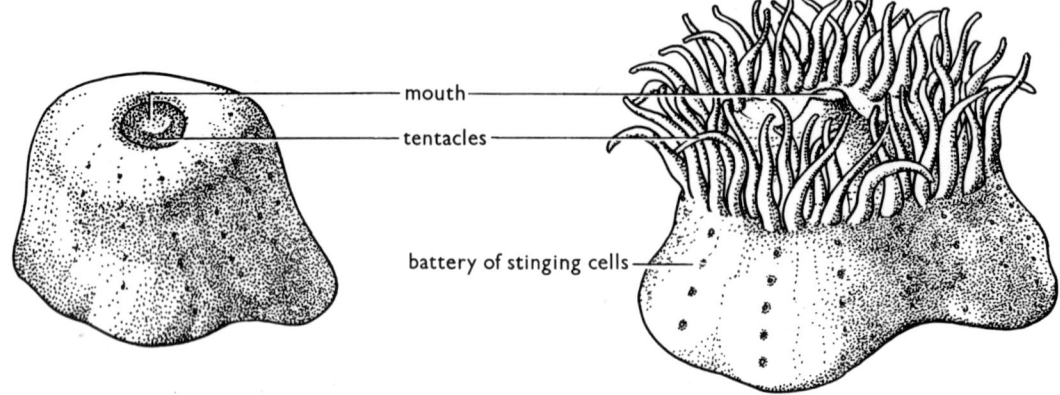

This is a common anemone found in rock pools on the sea shore. It is usually red in colour, and the body is a cylinder crowned by some 200 tentacles with a mouth at their centre. When touched the animal contracts to a compact mass about 3 cm in size, and the tentacles are withdrawn inside a ring of muscle. It can move very slowly on its muscular foot, by which it is normally attached to the rock. It feeds on small animals, which are paralysed when they collide with stinging cells on the anemone's body and tentacles. The prey is passed via the mouth into the digestive cavity by the tentacles. Any un-digested remains are ejected through the mouth.

The sexes are separate, and eggs and sperms are produced in reproductive organs inside the body. The fertilized eggs develop into small ciliated larvae which often remain inside the parent for some time before swimming away to form a new individual.

Examine an anemone in a small dish of sea water, and notice its reaction when touched gently and roughly. Observe it feeding on small pieces of chopped worm.

6.3 The Common Jelly Fish (*Aurelia aurita*)

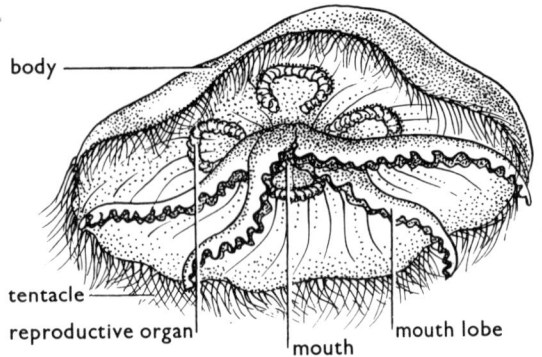

This Coelenterate lives floating in the sea, mainly in coastal waters. Its umbrella-shaped body is largely made from jelly and is colourless, reaching some 25 cm in diameter. Eight sense organs evenly spaced round the margin of the umbrella help the animal to keep right way up in the sea. It feeds on small animals in the plankton which are paralysed by stinging cells, then passed into the mouth by the action of cilia and movements of the mouth lobes. The sexes are separate, and there are four crescent-shaped reproductive organs, purple in colour.

If possible, examine a living jelly fish in a bucket of sea water and observe its method of movement. Alternatively examine the external structure of a preserved specimen.

6.4 By the Wind Sailor (*Vellela spirans*)

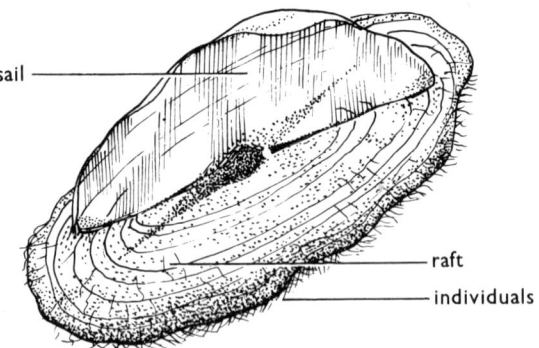

This animal has the appearance of a floating raft some 10 cm long with a diagonally set sail 6 cm high. It is blue in colour and floats on the surface of the sea. It originates in tropical waters and is often blown on to our south-west coasts by the prevailing winds. It is thought to be a colony of Hydra-like individuals which hang below the raft; a central one acts as the mouth, and of the rest—some are reproductive and others are provided with batteries of stinging cells.

Examine a preserved specimen and notice the texture of the raft and sail.

7 Platyhelminthes

Animals made from three layers of cells, ectoderm, mesoderm and endoderm (*Triploblastic*). *The gut when present has only one opening and there is no body cavity or* coelom.

7.1 Flatworm (*Planaria* spp.)

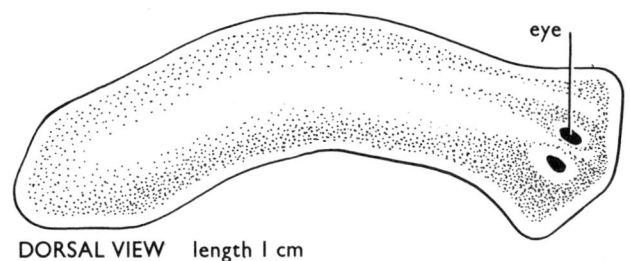

DORSAL VIEW   length 1 cm

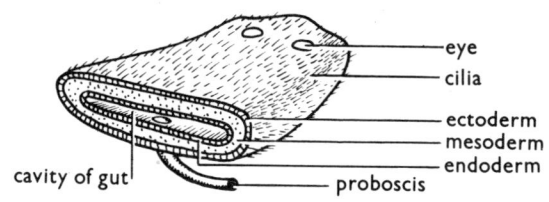

SECTION OF ANTERIOR HALF TO SHOW STRUCTURE

It is a black, bilaterally symmetrical, ribbon-like worm found under stones in fresh water. The head carries a pair of eyes and it moves by means of cilia covering the ectoderm. (See 18.1.) It is carnivorous, and has a short ventrally placed proboscis leading into a simple sack-like gut. The animal is hermaphrodite, and also has high powers of regeneration. (See 23.2.) Examine the animal with the lens or under the low power of the microscope in a drop of water in a watch-glass.

7.2 Beef Tapeworm (*Taenia saginata*)

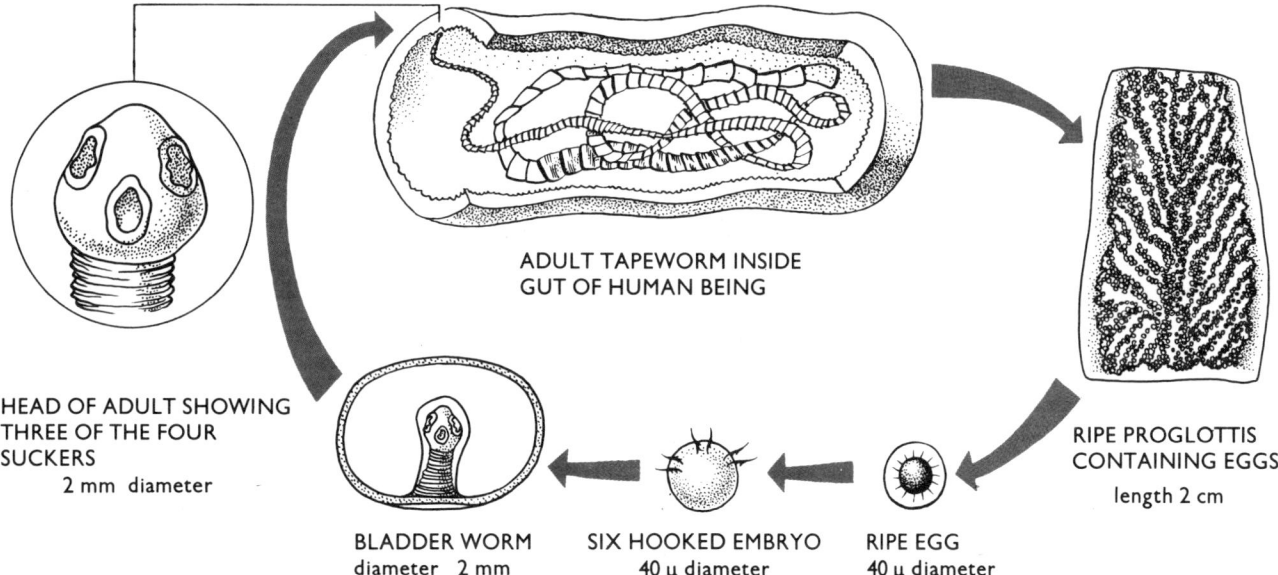

ADULT TAPEWORM INSIDE GUT OF HUMAN BEING

HEAD OF ADULT SHOWING THREE OF THE FOUR SUCKERS
2 mm diameter

BLADDER WORM
diameter 2 mm

SIX HOOKED EMBRYO
40 μ diameter

RIPE EGG
40 μ diameter

RIPE PROGLOTTIS CONTAINING EGGS
length 2 cm

Habitat The adult is parasitic in the gut of man, said to be the 'host'. Young tapeworm larvae inhabit the muscles of a cow.

Structure of the adult A flat ribbon-like worm several feet long. The head has 4 suckers fixing it to the wall of the host's gut, behind it a neck region continuously forms a series of *proglottides*, reaching 400 in number. The oldest ones farthest from the neck contain ripe eggs.

Irritability and movement There are no obvious sense organs, the nervous system is poorly developed. It can make feeble muscular movements.

Nutrition and respiration There is no gut. Food and air (probably in short supply) are absorbed over the surface of the *proglottides*.

Reproduction and life-cycle Each *proglottis* is hermaphrodite, and eggs are produced by self-fertilization within it. Ripe *proglottides* contain 40,000 eggs, and 10 *proglottides* may pass into the soil via the host's faeces each day. The eggs are set free, and if eaten by a cow, a small embryo with 6 hooks escapes from the egg into its gut. It burrows through the tissues, and reaches muscles in which it develops into a bladder with a small tapeworm head inside. If uncooked meat containing bladder worms is eaten by a human being, the small head turns inside out, and the 4 suckers attach it to the wall of the gut. A neck develops to form a new worm.

Method of examination Tapeworms are not common. It is best to study preserved specimens. The structure of the adult and its *proglottides* can be studied with a lens, or under the low power of the microscope.

7.3 The Liver Fluke (*Fasciola hepatica*)

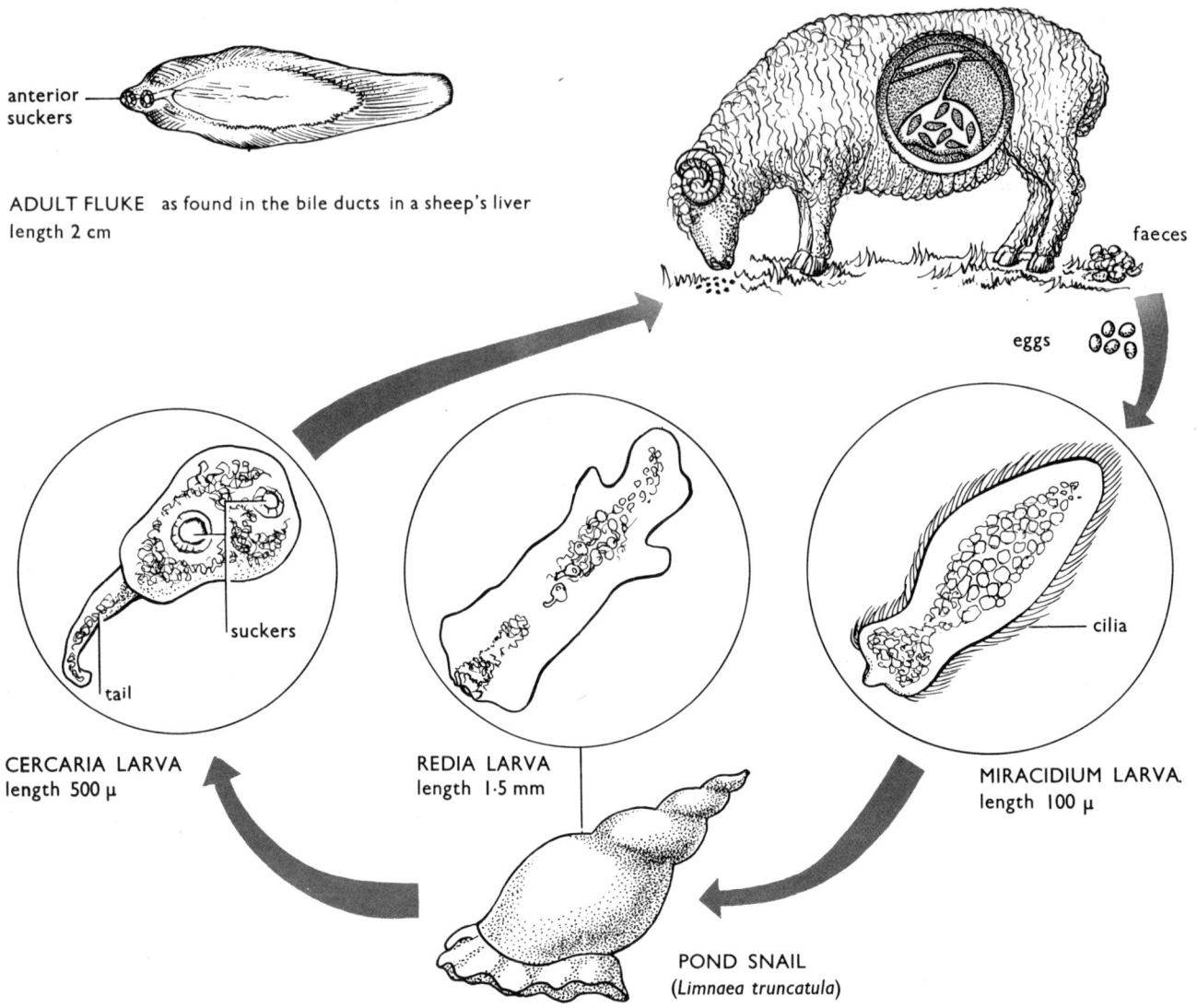

anterior suckers

ADULT FLUKE as found in the bile ducts in a sheep's liver
length 2 cm

faeces

eggs

CERCARIA LARVA
length 500 μ

suckers

tail

REDIA LARVA
length 1·5 mm

MIRACIDIUM LARVA
length 100 μ

cilia

POND SNAIL
(*Limnaea truncatula*)

Habitat of the adult   A parasite inhabiting the bile ducts of the sheep's liver.

External appearance of the adult   A flat leaf-like worm 2 cm long. Two suckers towards the anterior end serve to attach it to the lining of the bile ducts.

Irritability   There are no obvious sense organs, the nervous system is poorly developed.

Movement   The animal can make feeble wriggling movements.

Nutrition   A sack-like gut leads from the anteriorly placed mouth. Blood from the host is sucked in through the mouth, sugars can also be absorbed over the whole surface of the body.

Respiration   There are no special organs. Air obtained from the host's bile is absorbed over the surface of the body.

Excretion   Excretory substances are expelled through a pore at the posterior end of the body.

Reproduction and life-cycle   The animal is hermaphrodite, and is capable of producing several thousand eggs per day. These pass out in the host's faeces, and if they reach water, hatch out into a microscopic ciliated larva, a MIRACIDIUM. This burrows into a pond snail (*Limnaea truncatula*) forming a SPOROCYST, which quickly produces larvae termed REDIAE. These reproduce many times, and finally produce a third type of larva called a CERCARIA. This escapes from the snail, swims to the margin of the water, and forms a cyst. This is a resting stage, which, if eaten by a sheep, develops into a young Fluke which proceeds to the liver, and starts a new life-cycle.

Method of examination   Examine the adult animal with a hand lens, in a watch-glass placed on a dark background. Examine the tissues of a recently killed snail for larval stages. Examine prepared slides of larvae.

8 Nematoda

Cylindrical animals made from three layers of cells, ectoderm, mesoderm *and* endoderm (*triploblastic*). *The gut is a tube with an opening at the anterior and posterior ends.*

8.1 The Pig Roundworm (*Ascaris lumbricoides*)

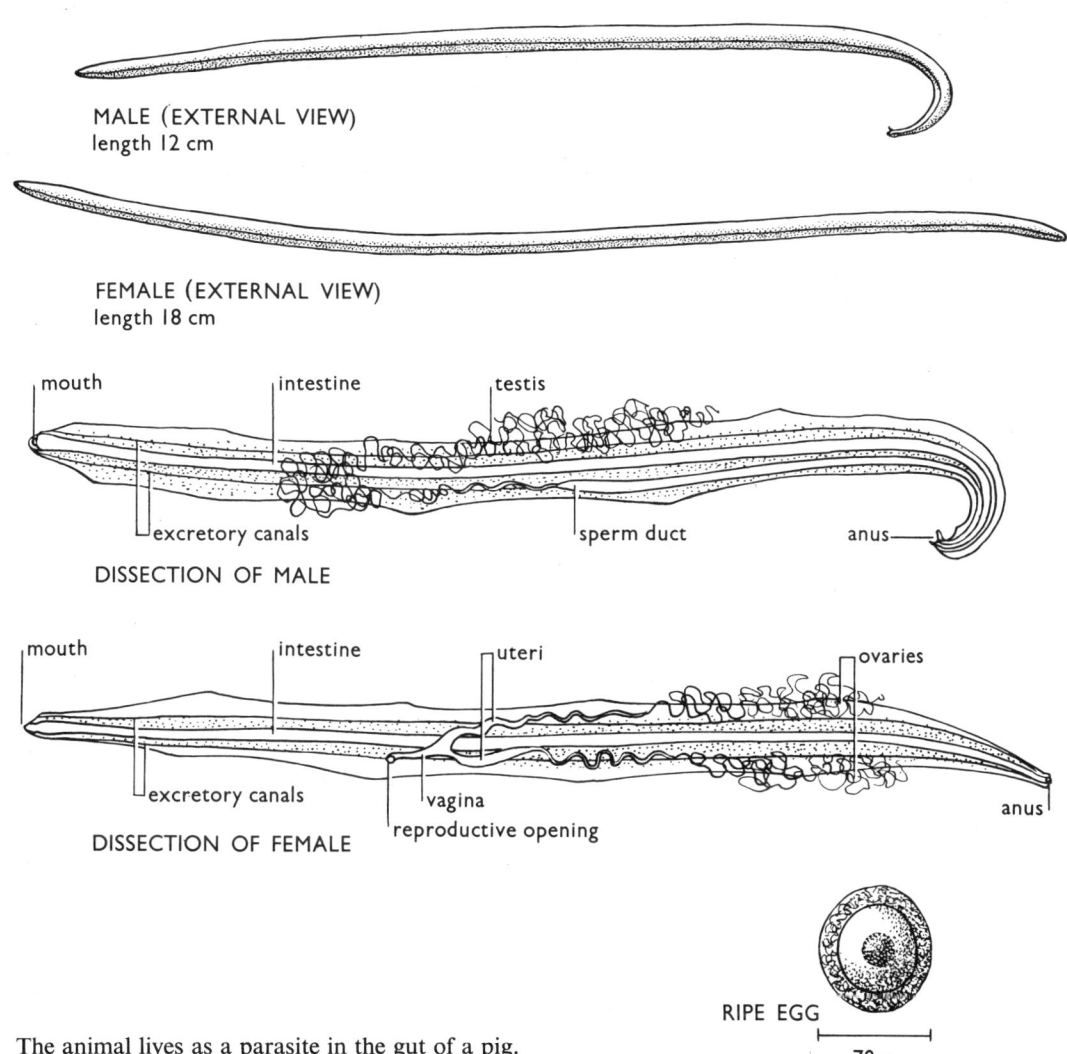

MALE (EXTERNAL VIEW)
length 12 cm

FEMALE (EXTERNAL VIEW)
length 18 cm

mouth   intestine   testis

excretory canals   sperm duct   anus

DISSECTION OF MALE

mouth   intestine   uteri   ovaries

excretory canals   vagina   anus
reproductive opening

DISSECTION OF FEMALE

RIPE EGG
70 μ

Habitat   The animal lives as a parasite in the gut of a pig.

External appearance   A creamy-white cylindrical animal with a shiny cuticle. A groove runs along the dorsal and ventral surfaces. Three lips round the mouth.

Irritability   There are no obvious sense organs, the nervous system is poorly developed.

Movement   The animal can make feeble wriggling movements.

Nutrition   The animal absorbs food which has been partly digested by its host.

Respiration   There are no special organs. Air (probably in short supply in its habitat) is absorbed over the surface of the body.

Excretion   A pair of canals along each side, open at a ventral pore at the anterior end of the body.

Reproduction   The male has one testis, connected by a sperm tube to the reproductive opening in its curled tail. The female has two ovaries, each connected by an *oviduct* and *uterus* to a single *vagina*, opening ventrally a short distance behind the mouth. Males and females copulate in the host's gut, and thousands of eggs pass out daily in the faeces. If eaten by another host (pig or man), they enter the gut and start another generation of roundworms.

Method of dissection   Pin the worm out under water in a dissecting dish, ventral surface towards the wax. Use a pair of scissors and forceps to open the body carefully from posterior to anterior. Identify the internal organs. Squash a portion of the uterus on a slide, and examine the eggs under the low power of the microscope.

17

9   Annelida

Cylindrical triploblastic animals with a gut opening at anterior and posterior ends. The body ringed externally and divided internally into about 150 segments. There is a body cavity or coelom.

9.1   The Earthworm (*Lumbricus terrestris*)

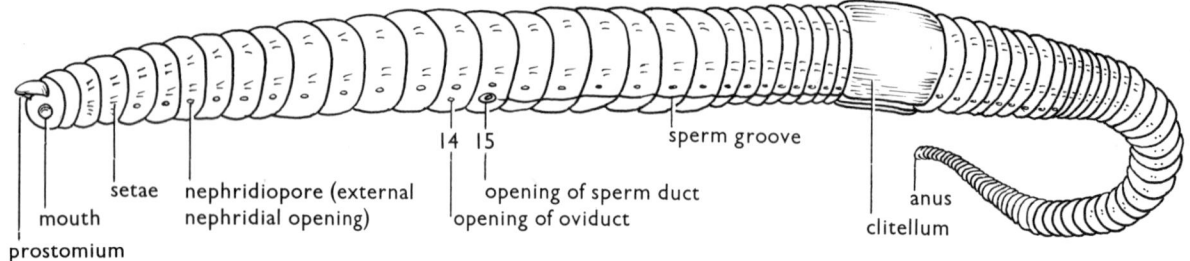

EXTERNAL APPEARANCE
FROM LEFT HAND SIDE

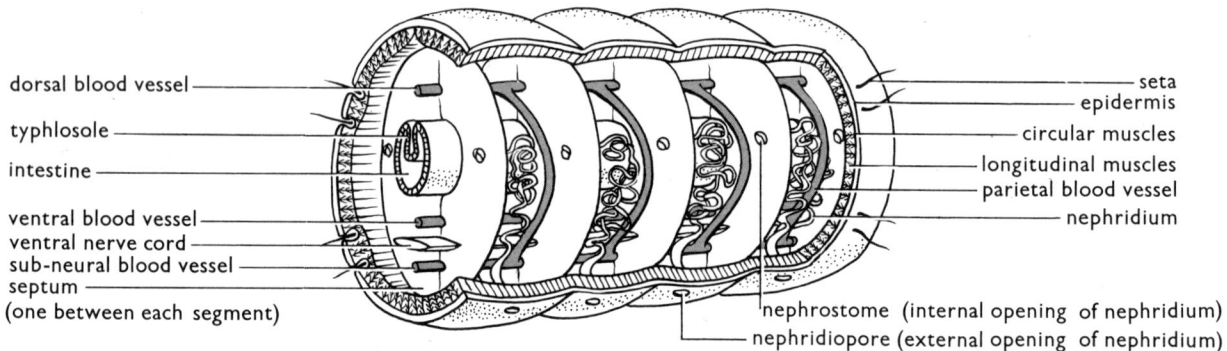

SECTIONAL DIAGRAM
SHOWING FOUR SEGMENTS
BEHIND THE CLITELLUM

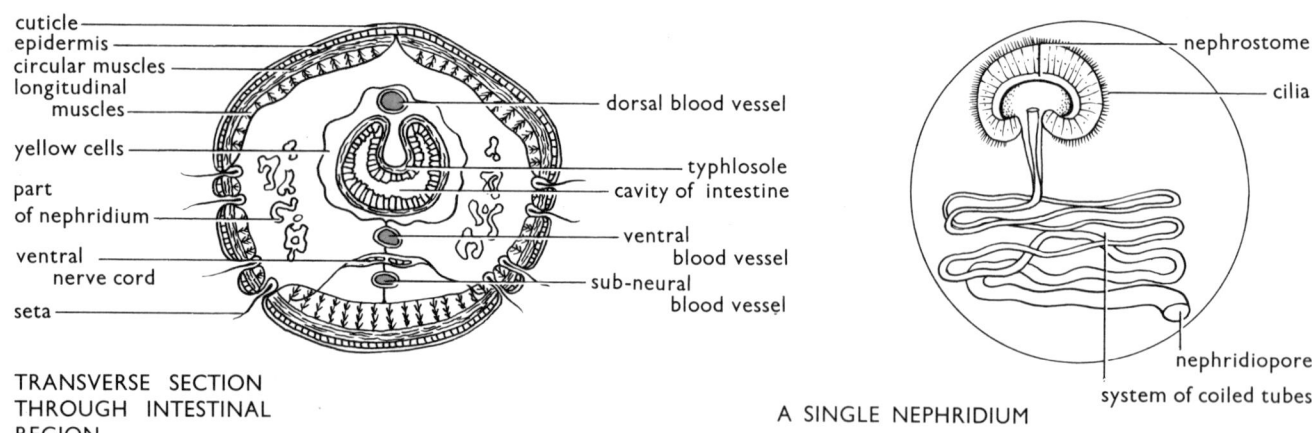

TRANSVERSE SECTION
THROUGH INTESTINAL
REGION

A SINGLE NEPHRIDIUM

Habitat and structure   It lives in burrows in the soil. It is cylindrical in form, often reaching 25 cm in length, the dorsal surface darker than the ventral. Each segment has 4 pairs of *setae* projecting laterally from it. There is a *clitellum* or *saddle*, one-third the distance from the anterior end.

Irritability and Movement   A ventral nerve cord extends the length of the animal below the gut. There is a ganglion in each segment, and a dorsally placed 'brain' in segment 3, joined to the rest of the nervous system by two *commissures* which surround the *pharynx*. There are sense organs in the skin. (See 18.1.) Alternate contractions of the muscles under the skin cause the body to elongate or shorten, the *setae* enable it to grip the sides of the burrows, enabling it to move forwards or backwards.

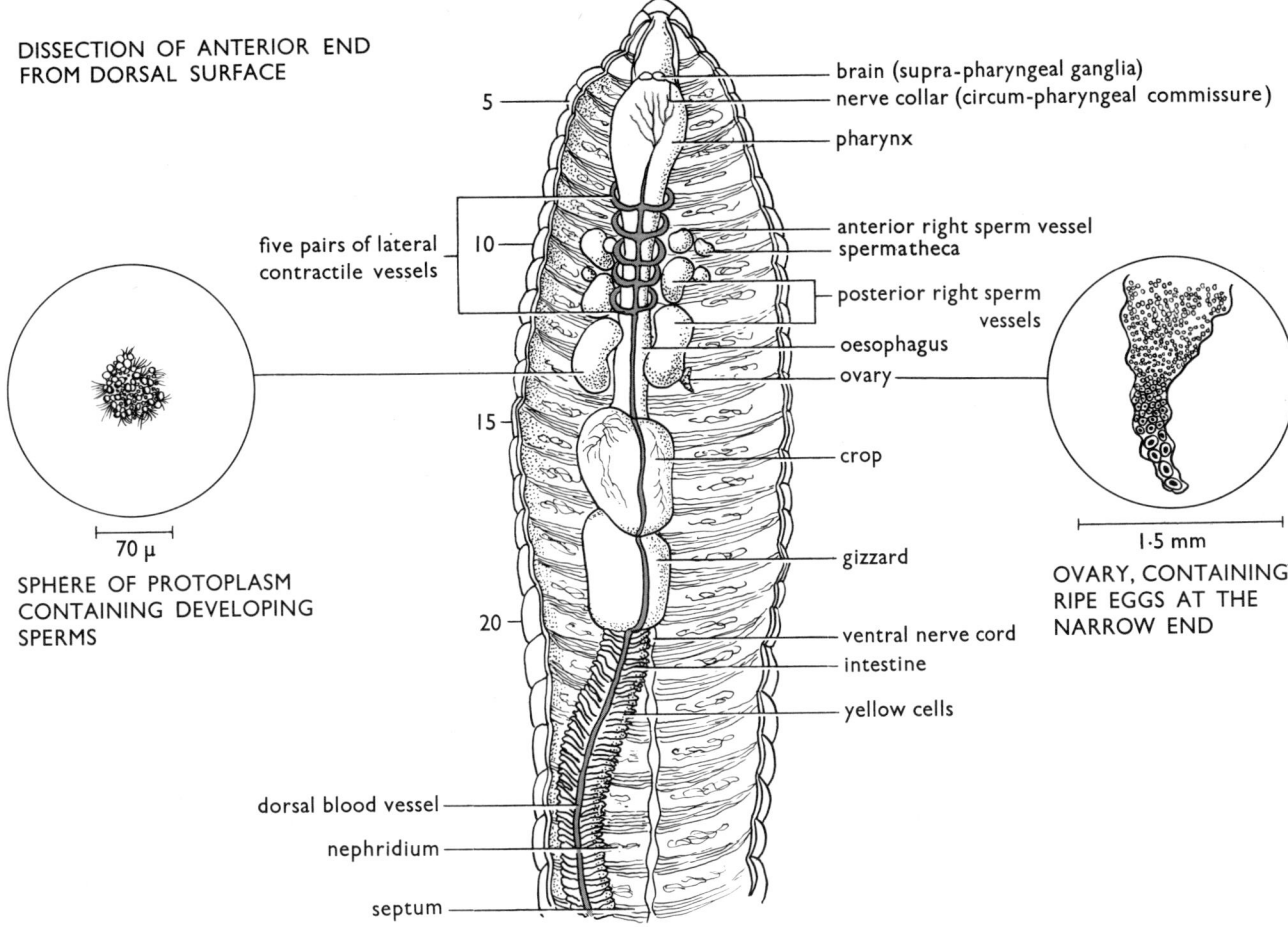

DISSECTION OF ANTERIOR END
FROM DORSAL SURFACE

5

five pairs of lateral
contractile vessels

10

15

20

brain (supra-pharyngeal ganglia)
nerve collar (circum-pharyngeal commissure)
pharynx
anterior right sperm vessel
spermatheca
posterior right sperm vessels
oesophagus
ovary
crop
gizzard
ventral nerve cord
intestine
yellow cells

dorsal blood vessel
nephridium
septum

70 μ

SPHERE OF PROTOPLASM
CONTAINING DEVELOPING
SPERMS

1·5 mm

OVARY, CONTAINING
RIPE EGGS AT THE
NARROW END

Nutrition The food is mainly leaves. Identify the main parts of the gut—*mouth* (segment 1), *buccal cavity* (1–2), *pharynx* (3–5), *oesophagus* (6–15), *crop* (16–17), *gizzard* (18–20). The rest of the body is occupied by the intestine, covered by a yellow tissue.

Respiration Gases diffuse through the skin. There is a blood system in which blood moves towards 5 pairs of hearts in segments 7–11. They surround the *oesophagus* and drive blood away in the ventral and sub-neural vessels. The blood contains the red pigment-*haemoglobin.*

Excretion A pair of ciliated coiled tubes called *nephridia* occurs in each of segments 4–149. They drain excretory materials through pores, a pair on the ventral surface of each segment.

Reproduction The animal is hermaphrodite. Identify the *spermatheca* in segments 9 and 10. The *sperm vessels* in segments 10–13. The pair of male openings ventrally placed on segment 15, and the female pair on segment 14. The *clitellum* from which the *cocoon* is made, segments 32–37.

Method of examination

(a) Examine a live earthworm, identify *prostomium, mouth, setae, male reproductive opening, sperm grooves, clitellum.* Observe the method of movement, identify dorsal and ventral surfaces.

(b) Dissect an earthworm, pinning its ventral surface downwards in a dissecting dish. Place two pins slantwise through the sides of segment 3, then, extending the animal, place another pin through the posterior end. Cover with water in which a pinch of salt has been dissolved. Make a cut in the dorsal surface near the posterior, with a pair of scissors. Cut forward carefully along the mid-dorsal line. Then carefully sever the septa along each side of the gut with a scalpel, paying special attention to the segments in front of the clitellum. Push a portion of the intestine to one side, note the ventral nerve cord beneath it.

(c) Remove a nephridium by snipping a septum and the coiled tube attached to it, then take it out with forceps. Mount in a drop of 1% salt solution on a slide, and note the rhythmic movement of the cilia when examined under the microscope. Nematode worms may also be seen.

(d) Squash a small portion of a sperm vessel on a slide, and smear its contents. Observe the sperms in various stages of development under low and high powers of the microscope. Stages of a Protozoan parasite called *Monocystis* may also be present.

(e) Examine a prepared slide of an ovary.

10 Arthropoda

Triploblastic animals, the coelom largely filled by organs, muscles and large spaces containing blood. There is a hard segmented external skeleton, flexible at the joints. Paired jointed appendages occur on most segments. Growth takes place by a series of moults.

10.1 CLASS CRUSTACEA

Mainly aquatic animals, with tough skeletons divided into approximately 20 segments.

10.1.1 The Crayfish (*Astacus fluviatilis*)

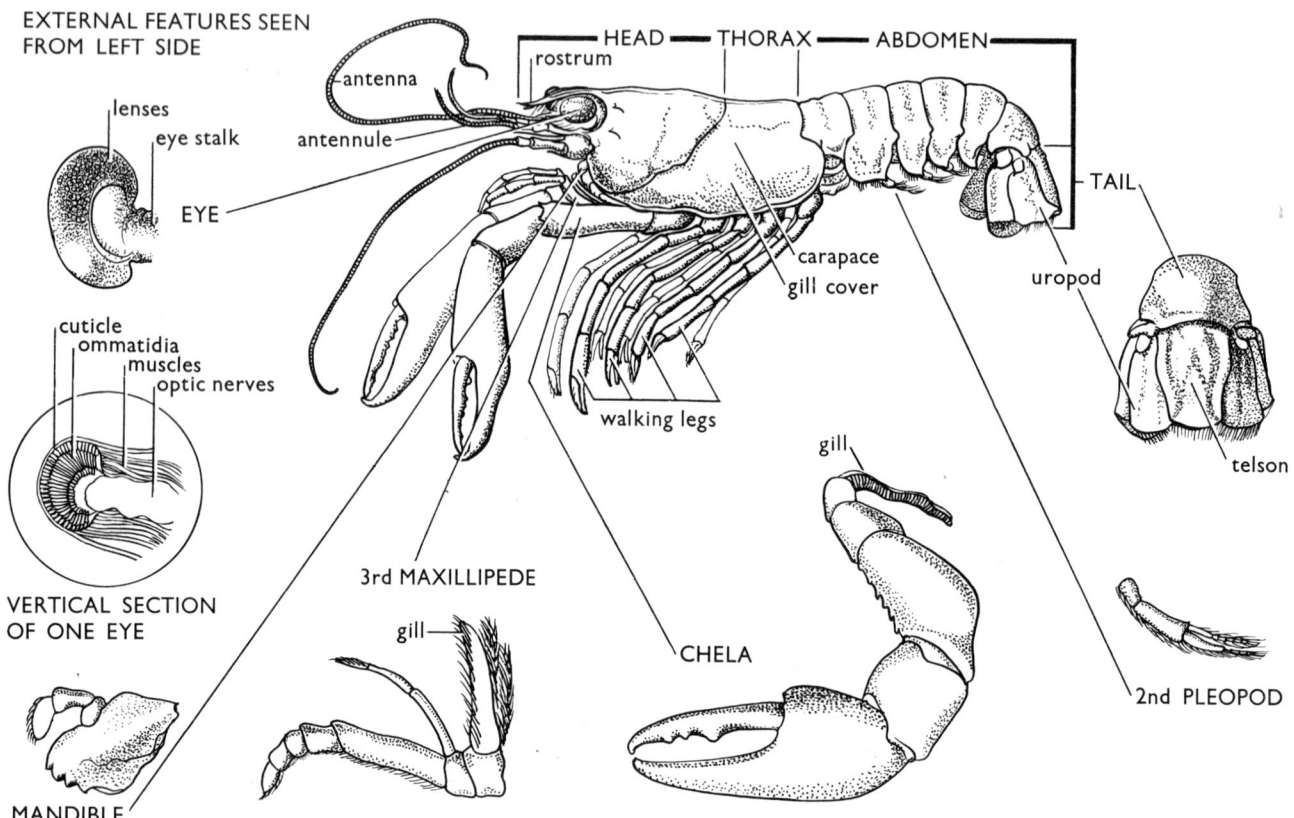

EXTERNAL FEATURES SEEN FROM LEFT SIDE

VERTICAL SECTION OF ONE EYE

MANDIBLE

3rd MAXILLIPEDE

CHELA

2nd PLEOPOD

Habitat and structure It lives in rivers and streams. The tough external skeleton is divided into head, thorax and abdomen, the latter having flexible segments. The paired appendages are arranged: head, 6; thorax, 8; abdomen, 6.

Irritability A ventral nerve cord runs the length of the body. The 'brain' in the head is joined to it by two nerves passing either side of the *oesophagus*. A pair of compound eyes and two pairs of antennae are the main sense organs found on the head.

Movement and nutrition It moves forwards on its thoracic walking legs, assisted by the abdominal swimmerets. When in danger it can make sudden movements backwards by flexing the abdomen and tail fan. It feeds on small animals, seizing them with its *chelae* and tearing with the *mandibles*. The prey is passed finally into the mouth, and ground and digested in the *gastric mill*.

Respiration and excretion It breathes through gills, many of which are joined to the thoracic appendages and situated under the thoracic gill covers. A current of water is drawn forwards over them, from behind; the oxygen they extract is carried round the body by a blood system. Excretory substances are removed by a pair of *green glands* in the head.

Reproduction The sexes are separate. The female lays eggs in autumn and carries them under her abdomen. When the larvae hatch out they stay attached to the swimmerets for a time. The crayfish also has the power to regenerate damaged limbs.

Method of examination

(*a*) Examine a live animal moving in a dish of water. Observe the mouth part which draws a current of water over the gills. Place a little ink from a pipette near the back of the thorax, and observe its passage through the front end of the gill chamber. Place one animal on a dark and another on a light background and notice how they adapt themselves by slowly changing colour.

(*b*) Pin a dead crayfish ventral side downwards in a dissecting dish. Cover it with water, then remove each segment by starting from the posterior end, making a cut on each side with a pair of scissors. Continue forwards on each side of the thorax near its dorsal surface, bringing the two cuts together behind the eyes. Remove the horny covering of the head and thorax and identify the internal organs.

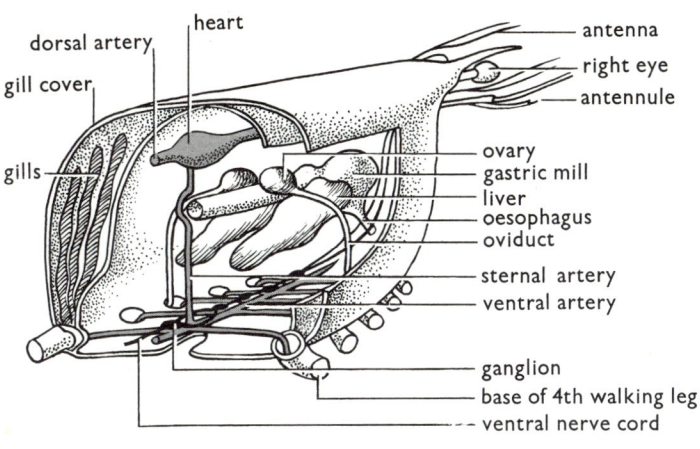

DIAGRAM FROM RIGHT
SIDE OF HEAD AND THORAX
TO SHOW INTERNAL
STRUCTURE OF FEMALE

dorsal artery
heart
gill cover
antenna
right eye
antennule
gills
ovary
gastric mill
liver
oesophagus
oviduct
sternal artery
ventral artery
ganglion
base of 4th walking leg
ventral nerve cord

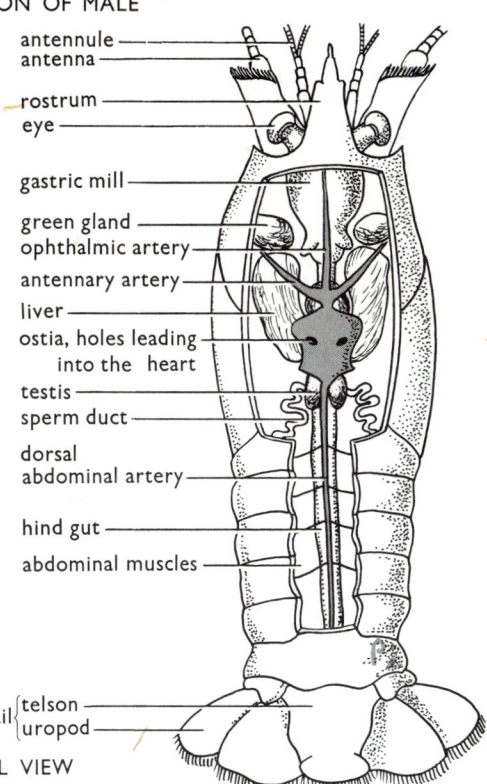

antennule
antenna
rostrum
eye
gastric mill
green gland
ophthalmic artery
antennary artery
liver
ostia, holes leading
    into the heart
testis
sperm duct
dorsal
abdominal artery
hind gut
abdominal muscles
tail {telson
     {uropod
DORSAL VIEW

OTHER CRUSTACEA

10.1.2 The Water Flea (*Daphnia pulex*)

10.1.3 The Garden Slater (*Oniscus asellus*)

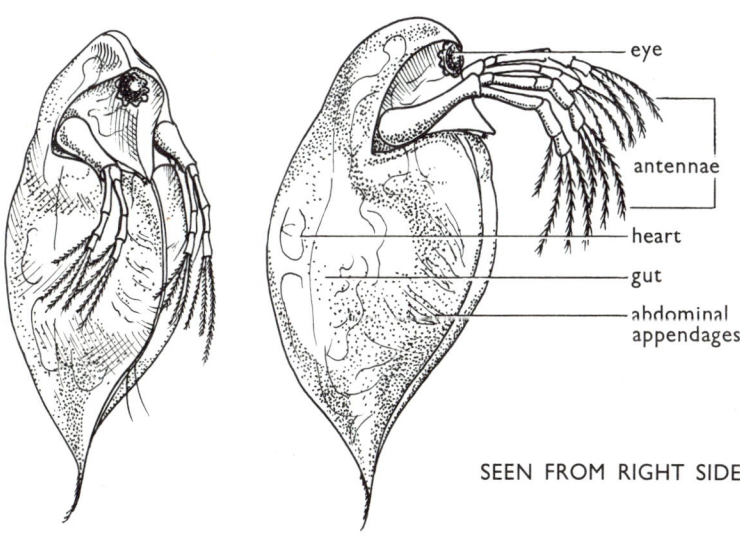

eye
antennae
heart
gut
abdominal appendages

SEEN FROM RIGHT SIDE

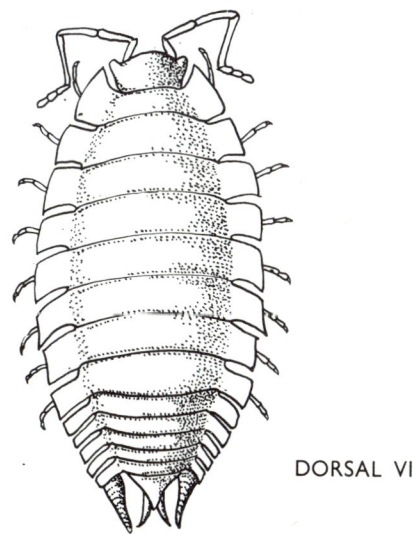

DORSAL VIEW

This is a very common Crustacean found in the surface waters of a pond or lake. It is 3–4 mm long. Examine it in a drop of water on a cavity microscope slide. Identify the pair of antennae which it uses for locomotion, and the prominent single eye formed from the fusion of two lateral eyes. The sides of the thorax are large and enclose the rest of the body and its appendages, which can be seen moving underneath it. Observe the heart in the dorsal region of the thorax. The rate at which it beats under various conditions of temperature can be studied. (See 18.2.)

This a common Crustacean often found under stones or leaves, and reaching a length of 14 mm. The segments of the thorax and abdomen are very similar in appearance, and most of their appendages are used in walking. The inner plates of the abdominal appendages function as gills; notice how the animal ventilates them by moving the abdomen up and down. These animals can be used for experiments to determine the reasons why they live in damp habitats. (See 18.0.)

10.2 CLASS INSECTA

Animals with a horny external skeleton, light in weight and divided into head, thorax and abdomen. The thorax has 3 pairs of legs placed ventrally, and there are 1 or 2 pairs of wings on the dorsal surface. The abdomen rarely carries any appendages.

10.2.1 The Cockroach (*Periplaneta americana*)

MALE (DORSAL VIEW)
body 5 cm long

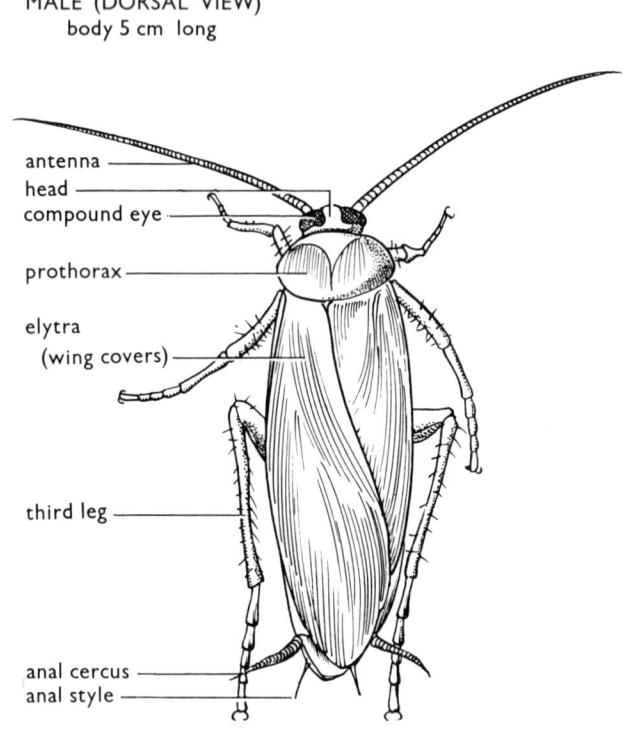

- antenna
- head
- compound eye
- prothorax
- elytra (wing covers)
- third leg
- anal cercus
- anal style

DISSECTION OF FEMALE
(dorsal view, head turned on to its right side)

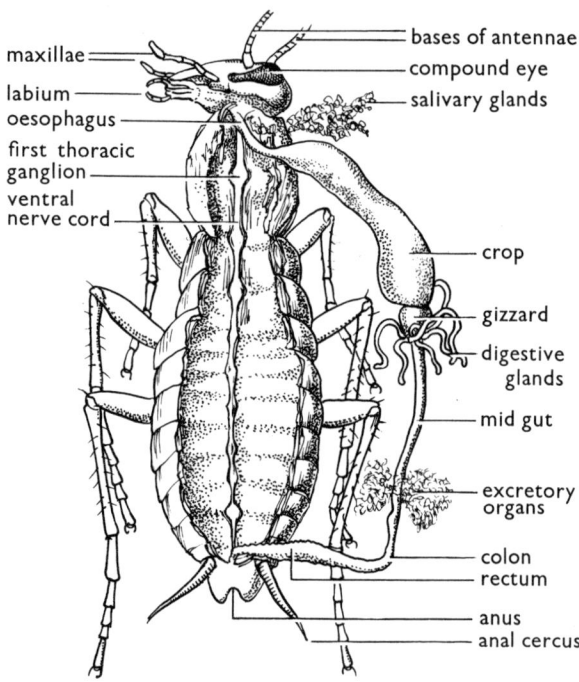

- maxillae
- labium
- oesophagus
- first thoracic ganglion
- ventral nerve cord
- bases of antennae
- compound eye
- salivary glands
- crop
- gizzard
- digestive glands
- mid gut
- excretory organs
- colon
- rectum
- anus
- anal cercus

Habitat and structure   It occurs in damp warm places such as cookhouses where it has become domesticated. Its natural habitat is tropical countries. The horny external skeleton is divided into head, thorax and abdomen. The head has a pair of long segmented antennae, a pair of compound eyes and 3 pairs of mouth parts. These are *mandibles, maxillae* and *labium*. There are 3 pairs of legs ventrally placed on the thorax; on the dorsal surface there is a pair of *wing covers* (*elytra*) attached to the middle segment, which protect delicate wings on the last thoracic segment. There are no abdominal appendages.

Irritability   There is a ventral nerve cord running the length of the body under the gut. A 'brain' placed dorsally in the head is connected to it by two nerves running on each side of the *oesophagus*. A pair of compound eyes fit closely to the front and sides of the head. There is a pair of long flexible antennae.

Movement   The insect probably never uses its wings for flying but instead walks actively on its legs. A pair of claws on the end of these enables it to climb vertical surfaces.

Nutrition   It feeds on scraps of food which it seizes and shreds with its mouth parts. Digestion is aided by saliva, produced from a pair of salivary glands placed on each side of the *oesophagus*.

Respiration   It breathes by means of tracheae, fine air tubes which carry oxygen to all parts of the body. Tracheae open to the exterior at holes called *spiracles*, 2 pairs on the sides of the thorax and 8 pairs on the sides of the abdomen. A large part of the body is filled with a colourless blood fluid, circulated by a thin tubular heart extending the length of the thorax and abdomen in the dorsal region.

Excretion   The excretory organs are fine tubes, joined to the hind portion of the midgut. They filter out waste materials from the blood cavity in which they lie.

22

DIAGRAM TO SHOW STRUCTURE
(SEEN FROM LEFT SIDE)

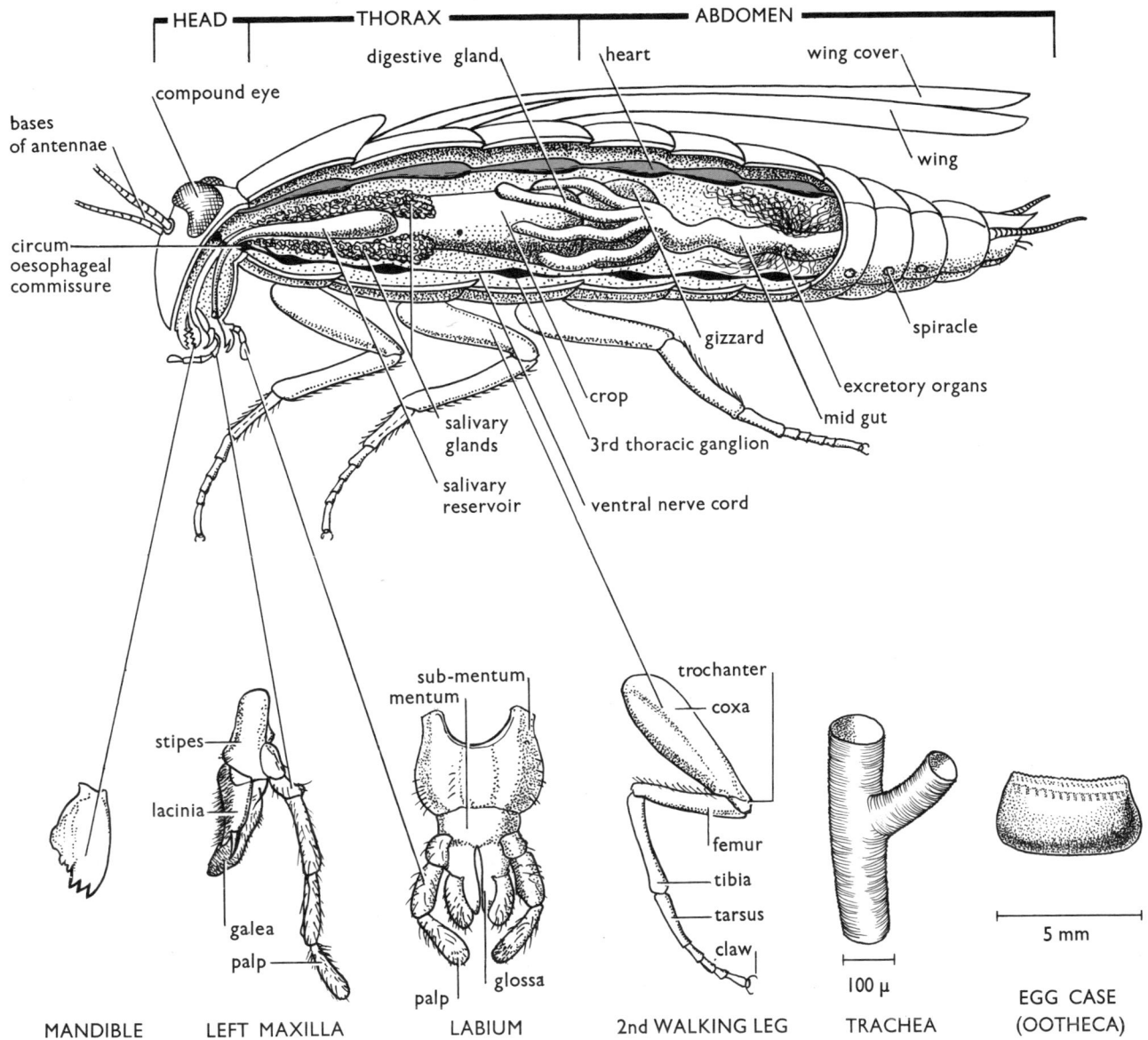

HEAD — THORAX — ABDOMEN

compound eye
digestive gland
heart
wing cover
bases of antennae
circum-oesophageal commissure
wing
salivary glands
salivary reservoir
crop
3rd thoracic ganglion
ventral nerve cord
gizzard
mid gut
excretory organs
spiracle

MANDIBLE
stipes
lacinia
galea
palp

LEFT MAXILLA
sub-mentum
mentum
palp
glossa

LABIUM

2nd WALKING LEG
trochanter
coxa
femur
tibia
tarsus
claw

TRACHEA
100 μ

EGG CASE (OOTHECA)
5 mm

Reproduction The sexes are separate. Males can be identified by 2 pairs of appendages at the posterior part of the abdomen, females have only 1 pair. The male injects sperms into the reproductive system of the female. The eggs when fertilized are laid 16 at a time and protected in a horny case which is carried under the abdomen of the female. The young cockroaches which hatch out are similar in structure to the adults, except that the wings are poorly developed; they are called *nymphs*, for they develop directly into the adult form by a series of moults.

Method of examination

(a) Examine living animals in a small covered glass dish. Notice how they move their antennae amd mouth parts, and also how they walk on the bottom and sides of the dish.

(b) Using a dead animal, carefully fix the head to the wax of a dissecting dish with a pin so that the mouth parts are towards you. Grip each part firmly with a pair of forceps and gently pull them away from the head. Mount the two *mandibles*, two *maxillae* and *labium* on a slide and examine their structure.

Dissect the animal by placing it in a small circular tin about 3 in. in diameter, containing wax. Melt the centre of the wax surface and place the animal abdominal side down, so that it sticks to it. Cover it with water. Remove the wing covers, and wings, then open the animal by cutting along each side of the abdomen forward to the head, using scissors and forceps. Remove the dorsal plates of the external skeleton carefully and notice the heart lying beneath it. Next remove this together with the fat body surrounding it, and examine and identify the remaining organs within the body.

Remove a piece of shiny *trachea*, and examine it in a drop of water on a microscope slide.

23

10.2.2 The Cabbage White Butterfly (*Pieris brassicae*)

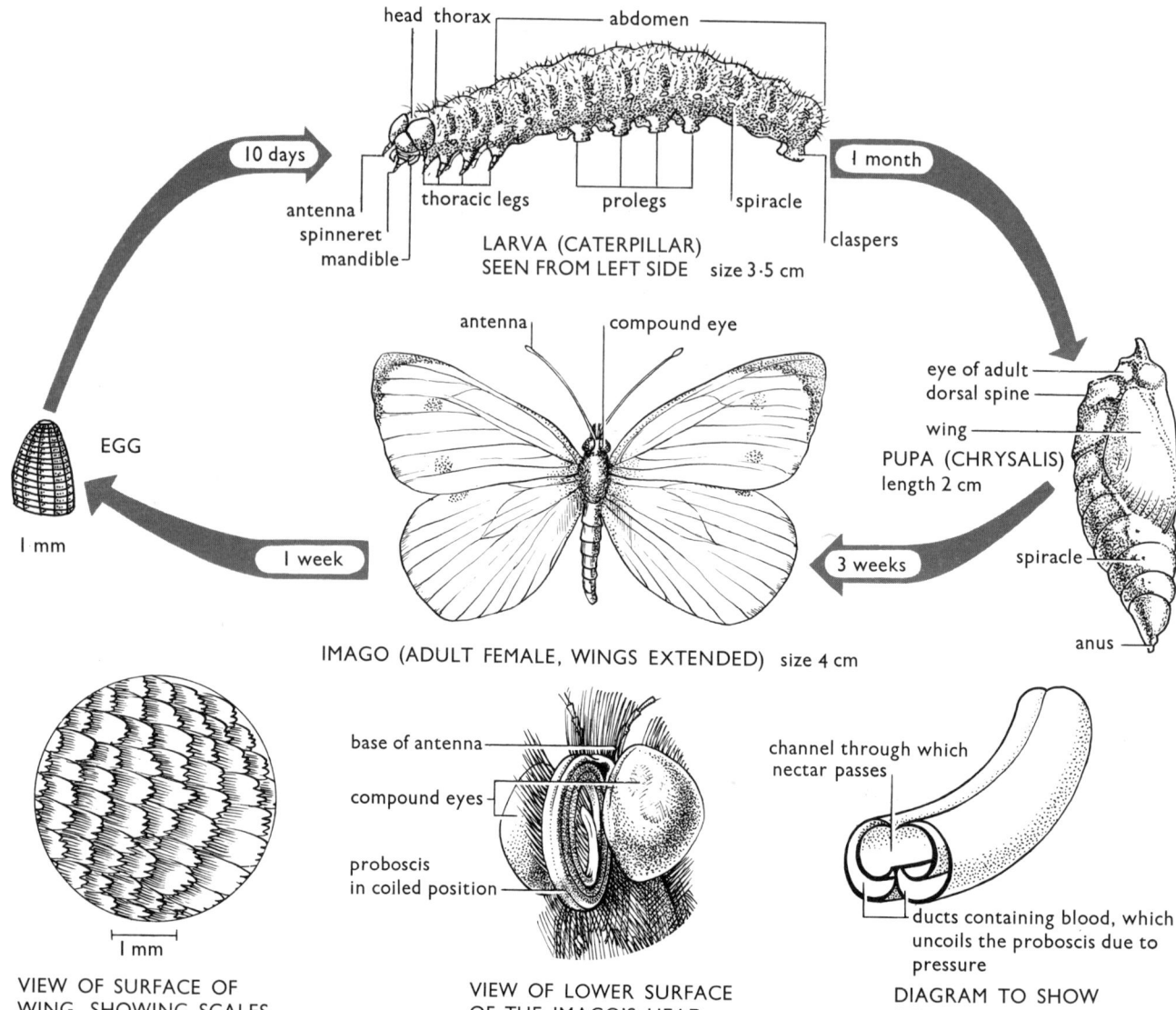

LARVA (CATERPILLAR) SEEN FROM LEFT SIDE size 3·5 cm

head thorax abdomen
antenna
spinneret
mandible
thoracic legs prolegs spiracle
claspers

10 days

1 month

EGG
1 mm

1 week

3 weeks

antenna compound eye

IMAGO (ADULT FEMALE, WINGS EXTENDED) size 4 cm

eye of adult
dorsal spine
wing
PUPA (CHRYSALIS) length 2 cm
spiracle
anus

VIEW OF SURFACE OF WING, SHOWING SCALES
1 mm

base of antenna
compound eyes
proboscis in coiled position
VIEW OF LOWER SURFACE OF THE IMAGO'S HEAD

channel through which nectar passes
ducts containing blood, which uncoils the proboscis due to pressure
DIAGRAM TO SHOW STRUCTURE OF PROBOSCIS

Habitat There are three distinct stages, the larva (or caterpillar) lives on cabbage leaves; the pupa (or chrysalis) in a dry, sheltered crevice; the adult is able to fly.

Structure of the adult It has a thin cylindrical body, clearly divided into head, thorax and abdomen. There are 2 pairs of wings, covered with fine scales. The fore wings of the female have two dark spots which are absent in the male.

Irritability The head carries a pair of long antennae, each one with a knob at the tip. There is also a pair of prominent compound eyes.

Movement It is rather clumsy in flight due to the feeble beating of the wings. When at rest they are folded in a vertical position.

Nutrition It feeds on nectar, which is sucked from flowers by the prominent *proboscis* situated under the head.

Respiration It breathes by means of *tracheae*, which open by *spiracles* on the sides of the abdomen (8 pairs) and thorax (1 pair).

Reproduction and life-cycle Eggs are laid by the female in batches under cabbage leaves, in early and late summer. They hatch out into caterpillars, cylindrical animals whose thorax and abdomen are somewhat similar in appearance. The head is distinct and carries a pair of small antennae, eye spots and a pair of powerful mandibles. The labium acts as a spinneret and produces silk. The whole of the body is hairy. After feeding voraciously and moulting about six times to reach full size, they move to a sheltered crevice into which they fix themselves with silk. They now enter the resting stage and develop into a pupa. During the course of a few weeks in summer (or over winter in the case of pupae produced in early autumn) the adult stage gradually develops and when mature emerges from the dorsal surface of the pupal case.

Method of examination Examine the behaviour of live caterpillars in a jar, containing cabbage leaves. If possible, allow them to pupate.

Examine preserved specimens of all three stages with a hand lens.

10.2.3 The Honey Bee (*Apis mellifica*)

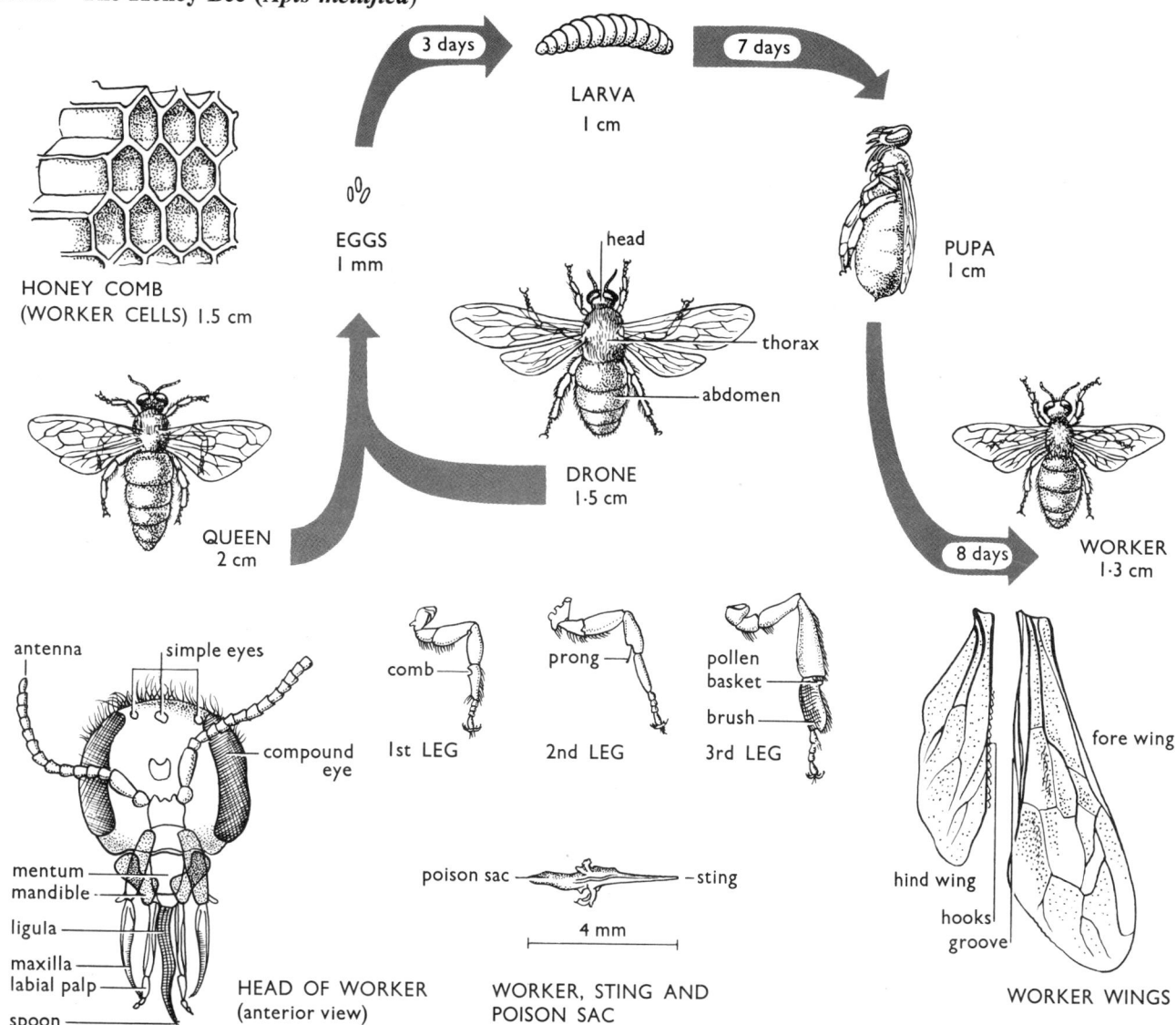

3 days

LARVA
1 cm

7 days

EGGS
1 mm

head

PUPA
1 cm

HONEY COMB
(WORKER CELLS) 1.5 cm

thorax

abdomen

QUEEN
2 cm

DRONE
1.5 cm

8 days

WORKER
1.3 cm

antenna   simple eyes

comb   prong   pollen
basket

brush

fore wing

compound
eye

1st LEG   2nd LEG   3rd LEG

mentum
mandible
ligula

maxilla
labial palp
spoon

HEAD OF WORKER
(anterior view)

poison sac   sting

4 mm

WORKER, STING AND
POISON SAC

hind wing

hooks
groove

WORKER WINGS

Habitat and structure It lives in a hive, a community consisting of a QUEEN (fertile female), several hundred DRONES (fertile males) and up to 70,000 WORKERS (sterile females). Each has head, thorax and abdomen, with 2 pairs of thoracic wings.

Irritability, movement and nutrition The head carries antennae, compound eyes and eye spots. The hind pair of wings carries hooks, locking in a groove on the fore wings. Bees feed on honey, made from pollen and nectar.

Reproduction and life-cycle A young princess receives sperms from a drone during the nuptial flight in May. She returns as queen of the hive and proceeds to lay about 1,000 eggs per day during the rest of the summer. Most of these are fertilized and develop into female larvae, but unfertilized eggs (about 1 in 8) develop into male drones. When the female larvae hatch out, the majority are fed on honey made from nectar, and these become workers. A few female larvae are fed on a special diet containing pollen grains, and these become future queens. The workers are the most numerous individuals and a timetable of their life cycle would be Day 0, egg laid. Day 3, larva hatches out. Day 10

pupation. Day 18, adult worker emerges. Day 18–28, the young workers make wax in glands under their abdomens and form it into honeycomb. They clean out cells. Day 28–32, they store food and ventilate the hive with their wings. Day 32–39 they make trial flights and do sentry duty. For a further 2 to 3 weeks they collect pollen and nectar. Pollen is carried on the body and in a pair of pollen baskets on the 3rd pair of legs. The 2nd pair have a prong for digging the pollen out on return to the hive. The 1st pair of legs carry a comb for cleaning the antennae. At the end of this period they die but workers still alive at the end of autumn survive the winter in the hive, together with the queen, feeding periodically on the honey store.

Any drones remaining in the hive at the end of the summer are pushed out by the workers and die of exposure.

Method of examination Examine live bees if possible in a demonstration glass-sided hive. Examine preserved specimens of the three castes in a watch glass, using a hand lens. Study prepared slides of worker mouth parts, legs, wings and sting. Study a piece of honeycomb, noting the brood cells, and the queen cells.

10.2.4 The House Fly (*Musca domestica*)

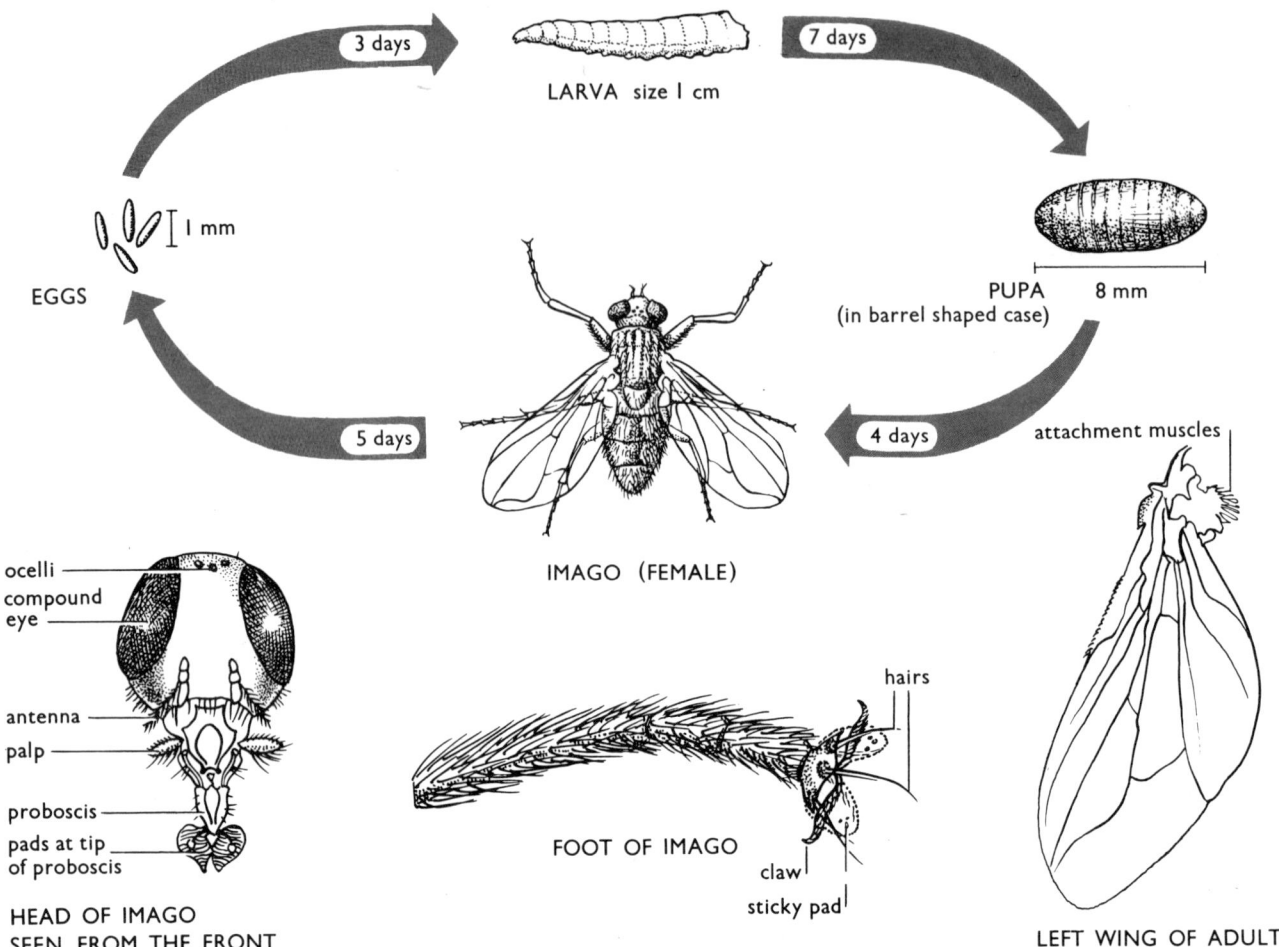

EGGS |1 mm

LARVA size 1 cm — 3 days

7 days

PUPA (in barrel shaped case) — 8 mm

4 days

IMAGO (FEMALE)

5 days

HEAD OF IMAGO SEEN FROM THE FRONT
- ocelli
- compound eye
- antenna
- palp
- proboscis
- pads at tip of proboscis

FOOT OF IMAGO
- hairs
- claw
- sticky pad

LEFT WING OF ADULT
- attachment muscles

Habitat There are three distinct stages. The larva or grub lives in rotting food, the pupa lives in a dry sheltered habitat, and the adult flies especially in and about human dwellings.

Structure of the adult There is a distinct head, thorax and abdomen. One pair of wings is situated on the middle thoracic segment. A pair of projections from the dorsal surface of the third thoracic segment is thought to act as a balancing mechanism when flying. The whole body is hairy.

Irritability There is a ventral nerve cord. The head has three simple eye spots placed anteriorly and a pair of compound eyes.

Movement It flies efficiently. Each pair of legs has 2 claws and 2 sticky pads at the tip. The pads enable the fly to walk upside down on smooth surfaces.

Nutrition It feeds on liquid food. The salivary glands produce a digestive fluid which is passed to the tip of the *proboscis*. This dissolves any food material on which the fly has settled, so enabling the fly to suck it up into its digestive system.

Respiration The fly breathes through a system of *tracheae* opening to the outside of the body by means of *spiracles*.

Reproduction and life-cycle The female lays small cylindrical eggs about 1 mm long. They are placed during the summer in batches in decaying food and hatch in a few days into larvae. These grow rapidly, moulting twice to reach a length of about 12 mm. They have 12 visible segments and are conical in shape, with the head at the pointed end. This carries 2 eye spots which are kept in the shadow of the bulkier body, so encouraging the larva to move away from light. Having reached full size in 7 days, it moves to a dry place where the skin becomes hardened and contracts into a barrel shape, brown in colour. Inside this a pupa develops. During this stage the tissues are reorganized, and in about 4 days an adult fly is formed. The lid of the pupa case is pushed off by the pressure of a small sac on the head which is inflated for this purpose. It is possible that in cold weather the pupal stage lasts for a much longer time, and it is in this condition that flies survive the winter. The adult fly is responsible for many human diseases due to its habits of feeding on manure and decaying food. Germs which it picks up on its hairy body are then transferred to any human food on which it may happen to alight.

Method of examination Examine the living larva of a bluebottle in a small dish. Observe its method of movement and reaction to light. (See 18.1.)

Examine a dead house fly with a hand lens. With a pair of forceps remove a wing, a leg and the *proboscis* and observe these on a slide under the low power of the microscope.

10.2.5 The Gnat (*Culex pipiens*)

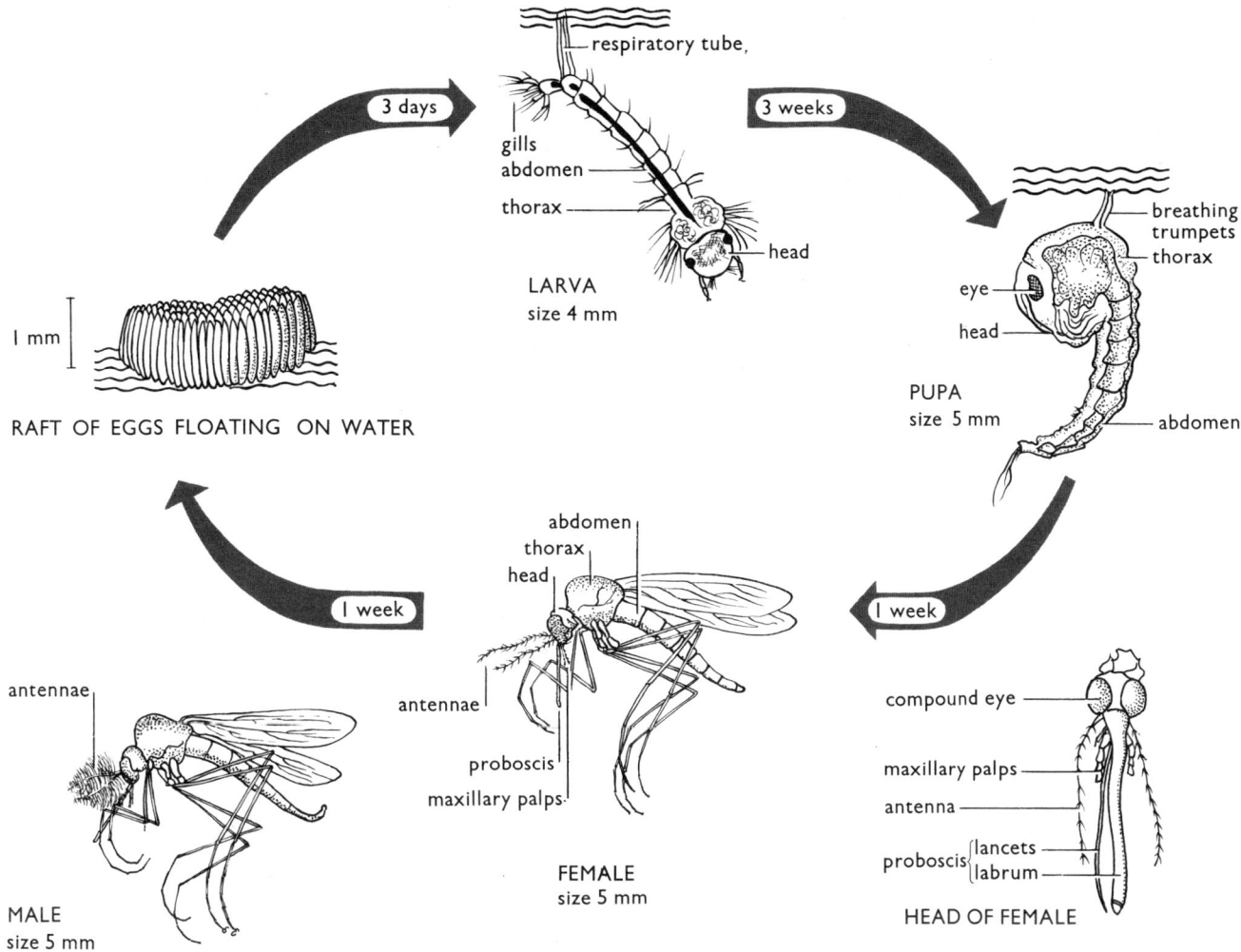

RAFT OF EGGS FLOATING ON WATER

LARVA
size 4 mm

PUPA
size 5 mm

MALE
size 5 mm

FEMALE
size 5 mm

HEAD OF FEMALE

Habitat   There are three distinct stages. The larva and pupa live in ponds and ditches, the adult is able to fly efficiently.

Structure of the adult   The body is divided into head, thorax and abdomen with a pair of wings on the middle thoracic segment. The male has short, hairy antennae and his mouth parts are adapted for sucking liquid food. The female has long thread-like antennae, and her mouth parts are adapted for both biting and sucking.

Irritability   There is a ventral nerve cord. The head carries a pair of large compound eyes.

Movement   The adults are efficient in flight, the beating of the wings making a low buzzing noise. The legs are slender.

Nutrition   The male feeds on nectar, sucked up into the digestive system by a *proboscis*. The female has mouth parts adapted for biting in addition, which can pierce human skin so that the insect can feed on blood. Saliva produced in the insect's salivary glands prevents the blood from clotting as it is sucked up by the *proboscis*. It also contains a substance which irritates the victim's skin.

Respiration   The insect breathes by a system of *tracheae* which opens to the exterior by means of *spiracles*.

Reproduction and life-cycle   The female lays the eggs in the form of rafts which float at the surface of still water, in ditches and small ponds. The larva hatches out in 3 days and is similar to a caterpillar in appearance; it can swim in the water. Periodically it comes to the surface in a horizontal position to breathe in air through a tube on the 8th abdominal segment. On the 9th segment there are gills. After moulting four times it changes into a pupa which continues as an active animal. The head and thorax are comparatively large, the abdomen being thin and curved underneath them. Two small breathing tubes project from the dorsal surface of the thorax and enable the animal to breathe air through the surface film of the water. It can swim up and down in the water by jerking movements of the abdomen. After a few days, the pupa comes to the surface of the water, its skeleton splits along the dorsal surface and the adult emerges. It rests for a time on the floating pupa case, until its wings harden, then it flies away.

Method of examination   Examine living larvae and pupae in a jar of water and observe their methods of movement.

Under the low power of the microscope examine prepared slides of eggs, larvae, pupae and adults. Compare the structure of the head appendages of adult males and females.

27

11 Mollusca

Triploblastic coelomate animals with fleshy unsegmented bodies and no appendages. A soft skin called the mantle secretes a chalky shell.

11.1 The Common Mussel (*Mytilus edulis*)

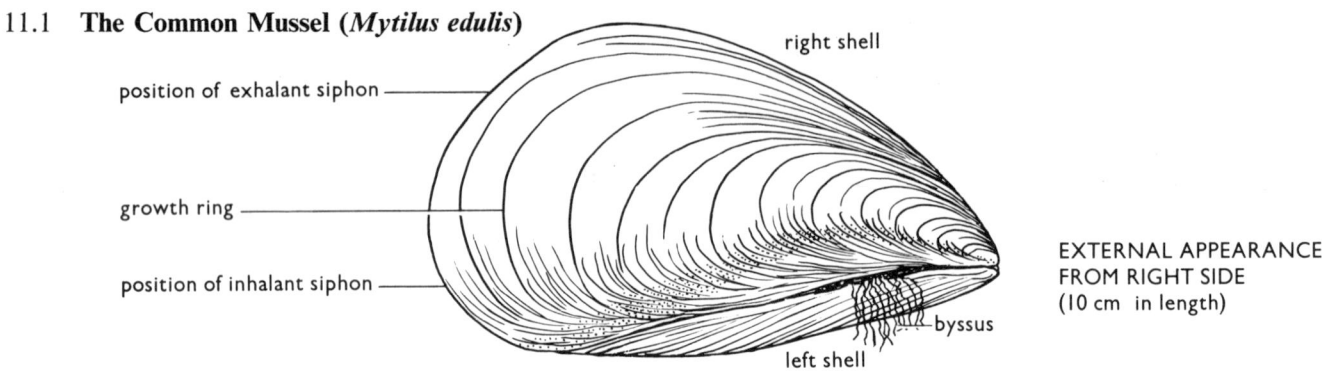

position of exhalant siphon

growth ring

position of inhalant siphon

right shell

left shell

byssus

EXTERNAL APPEARANCE
FROM RIGHT SIDE
(10 cm in length)

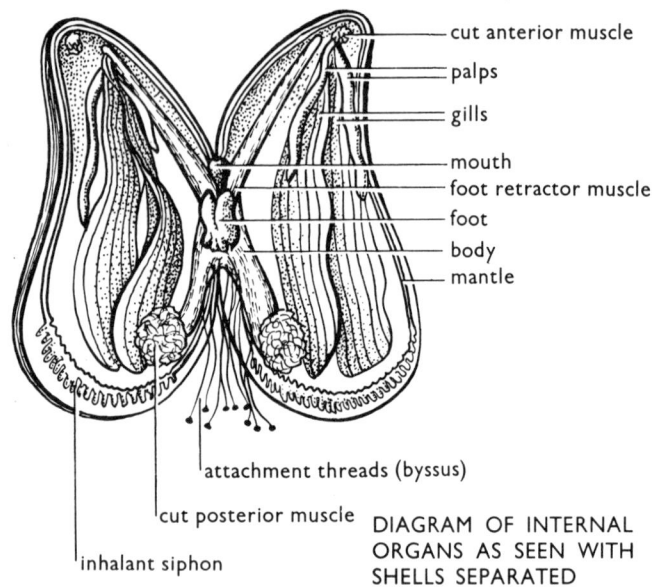

cut anterior muscle
palps
gills
mouth
foot retractor muscle
foot
body
mantle

attachment threads (byssus)

cut posterior muscle

inhalant siphon

DIAGRAM OF INTERNAL
ORGANS AS SEEN WITH
SHELLS SEPARATED

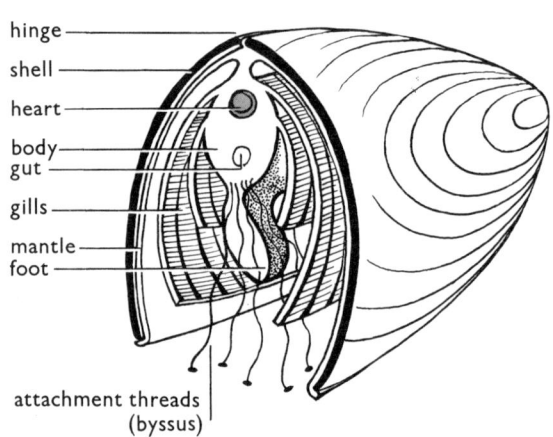

hinge
shell
heart
body
gut
gills
mantle
foot

attachment threads
(byssus)

DIAGRAMMATIC HALF SECTION

Habitat Marine animals, which live attached to rocks or piers near the low tide mark of coastal waters.

Structure The fleshy body is enclosed by two shells hinged at the apex of the animal. The shells are chalky in texture and are made by the mantle, which secretes them from its outer edge. As the mantle's rate of activity varies growth rings are visible on the outside of each shell.

Irritability There are no obvious sense organs and the nervous system is poorly developed.

Movement There is a flexible fleshy foot which can be protruded from the ventral surface between the shells when they are open. The animal can make slow movements. When disturbed the foot is withdrawn into the shells by powerful muscles and the shells themselves are closed by strong transverse muscles, one at the anterior and the other at the posterior end. Behind the foot, a gland makes horny threads which attach the animal to a rock so that it cannot be washed away by waves.

Nutrition It feeds on bacteria. These are swept into the space between the shells by a current of water caused by the rhythmic beating of cilia covering 2 pairs of gills which hang down on either side of the body. Water enters through an *inhalant siphon* at the posterior blunt end of the shells, and leaves through an *exhalant siphon* situated above it. Two pairs of anteriorly placed *palps* at each side of the mouth filter particles of food off the gills and pass them into the mouth.

Respiration The gills are also used for absorbing oxygen from the circulating current of water. There is a simple heart and blood system.

Excretion There is a kidney in the dorsal part of the body. Excretory materials are passed out through the exhalant siphon.

Reproduction The sexes are separate. Gametes from the ovaries and testes are shed into the sea where fertilization takes place. The eggs then develop into a small ciliated larva shaped like a spinning-top. If this is washed ashore by the waves, it settles on a rock and develops into the adult form.

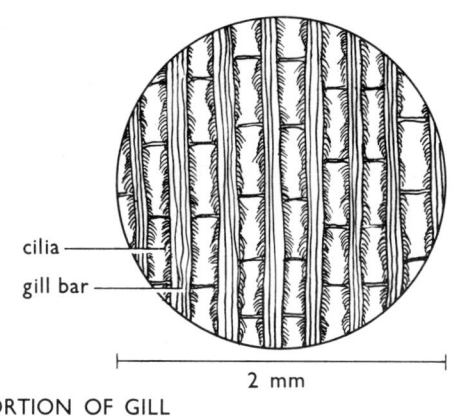

cilia

gill bar

2 mm

PORTION OF GILL
MAGNIFIED

Method of examination Place a living specimen in a small dish of sea water. With a pipette introduce a little carmine or ink close to the inhalant siphon. Observe the exhalant siphon and watch for the coloured particles emerging from it. If a living animal is placed in cloudy sea water, it will filter particles out of the water, so that it becomes clear in about 12 hr.

As the irritable system is poorly developed the animal can be opened alive. Insert a coin into the ventral region of the shells and prize them apart. The anterior and posterior muscles will be torn as the shells are parted. Identify the internal organs, by placing the animal in a dissecting dish containing sea water, so that the organs are supported by the sea water. Notice the irritability of the foot and palps when touched by a seeker. Cut out a small square of gill and mount it in a drop of sea water on a slide. Under the low power of the microscope observe the beating of the cilia. (See 18.2.)

Chip off a small piece of shell, and place it in a watch-glass containing dilute hydrochloric acid. Notice the evolution of carbon dioxide as the acid attacks the chalky material of the shell.

11.2 The Common Pond Snail (*Limnaea stagnalis*)

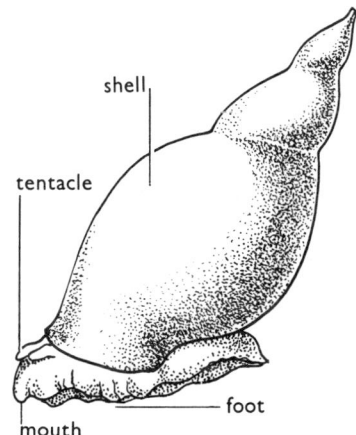

shell

tentacle

foot

mouth

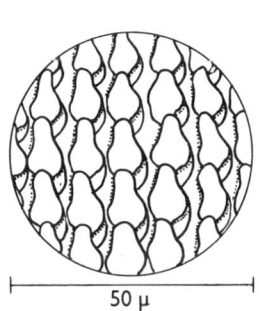

50 μ

PORTION OF RADULA

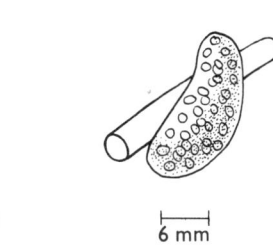

6 mm

EGGS IN CASE ON A
WATER PLANT STEM

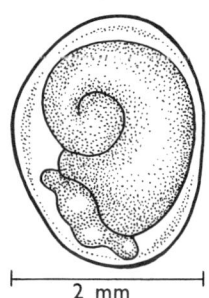

2 mm

YOUNG SNAIL
DEVELOPING
INSIDE EGG SHELL

Habitat It lives amongst vegetation in ponds and streams.

Structure It has one shell 3 cm long, and arranged in a spiral of approximately 6 turns rotating clockwise when viewed from the pointed end. It is made by a *mantle* which lines it. The fleshy body is protected by the shell.

Irritability There is a well-developed nervous system and the whole of the body is sensitive to touch. A pair of flexible tentacles on the head have a simple eye at the base of each one.

Movement The snail glides on its muscular foot and is able to move upside down on a surface film of water.

Nutrition It feeds on water plants. It rasps off pieces of leaf with a horny tongue (*radula*) situated in the lower part of the mouth. The radula is equipped with 500 rows of small teeth.

Respiration and excretion It breathes gaseous air by means of a lung lined with blood vessels situated between the mantle and the body. Oxygen is carried round the body in a blood system circulated by a powerful heart. The animal comes to the surface periodically to exchange stale air in the lungs for fresh air. Excretory substances are filtered from the blood by a kidney.

Reproduction The animal is hermaphrodite, but cross-fertilization takes place, the sperms from one animal fertilizing the eggs of another. The eggs are laid during the summer in a gelatinous mass each containing about 30 eggs. They hatch in approximately 1 month.

Method of examination Observe a snail in a small aquarium with a hand lens, and examine it moving on the glass. Watch for its *radula* to emerge and rasp off green plants on the side of the tank. Observe it surfacing in order to breathe, sometimes it does this by letting go so that the buoyancy of the air in its lung causes it to rise to the water surface rapidly.

Mount a snail egg on a slide, and examine it under the low power; the young developing snail can often be seen through the egg shell.

Examine a prepared slide of a radula.

29

12 Echinodermata

Radially symmetrical triploblastic coelomate animals covered with chalky spines. They are marine.

12.1 The Common Starfish (*Asterias rubens*)

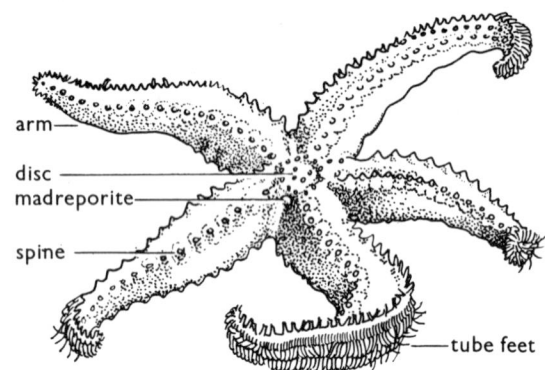

UPPER (ABORAL) SURFACE size 25 cm

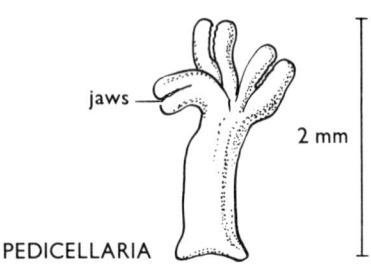

PEDICELLARIA

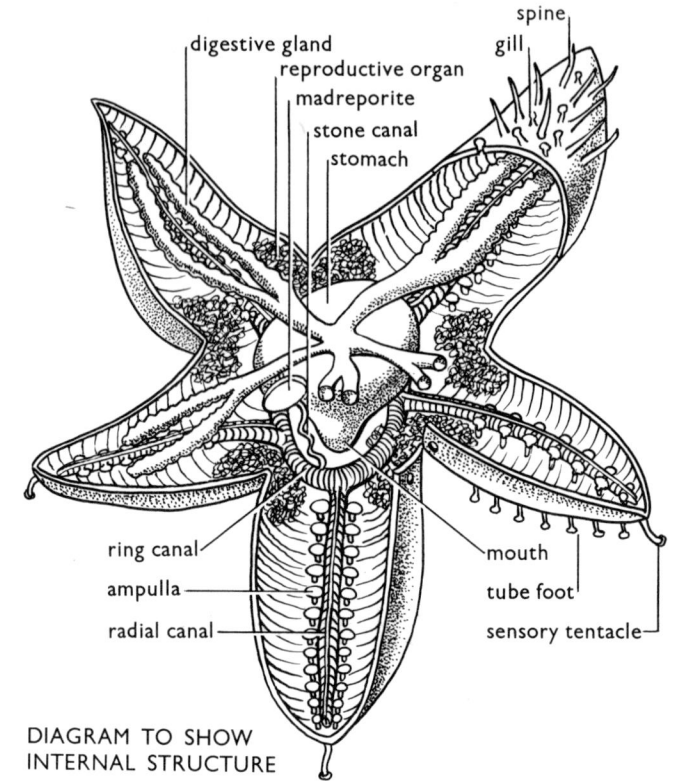

DIAGRAM TO SHOW
INTERNAL STRUCTURE

Habitat and structure It lives in shallow coastal waters and rock pools. It has a mouth on the lower (oral) surface of a central disc, from which 5 arms radiate. On their lower surface a system of tube feet runs in a groove. The whole upper surface (aboral) is covered with chalky spines amongst which are organs called *pedicellariae*. They keep the skin clean by rotating about their fixed bases and nipping off any foreign bodies with a pair of beak-like jaws.

Irritability and movement The nervous system is poorly developed. The main sense organs consist of a special tube foot at the tip of each arm. Locomotion is brought about by a system in which sea water enters a porous plate (the *madreporite*) asymmetrically placed on the aboral disc surface. Water is moved by cilia acting inside the pores and passes via a *stone canal* to a *ring canal* surrounding the mouth. From here, the continuous pressure caused by the cilia forces it into *radial canals* running in the lower part of each arm. The water pressure inflates small sacs called *ampullae* communicating with the tube feet. These extend and slow waving movements cause the animal to move, usually with one arm leading and the other four trailing behind.

Nutrition It feeds on mussels by placing its five arms over the two shells of its prey, slowly prising them apart, then through its mouth turns its stomach inside out directly on to the tissues of the mussel.

Respiration and excretion The hard aboral surface of the skin is pierced in many places by thin walled sacs which act as gills and excretory organs.

Reproduction The testes in the male and ovaries in the female lie in the angle of each arm. The gametes escape from an opening, one between each arm and rise to the surface of the sea. Here the eggs are fertilized and develop into ciliated larvae which feed on surface plankton. When fully grown (about 2 mm) they sink to the bottom and develop into adult starfish.

The starfish has good powers of regeneration.

Method of examination Observe live starfish in a dish of sea water, noting their method of movement. Place one upside down in the water and observe how it rights itself. In early summer it is sometimes possible to observe fertilization of starfish eggs. (See 23.3.)

Dissect a dead starfish, by placing its oral surface downwards in a dissecting dish, and place a pin through the tip of each arm. Cover it with water containing a pinch of salt. Cut round the edge of the disc, leaving the *madreporite* in position, and remove the aboral section of the disc. Cut along the aboral surface of each arm and pin the skin back. Identify the internal organs.

Cut off an arm close to the disc and insert a fine pipette containing water into the exposed radial canal of the water vascular system. Squeeze water from the pipette into this canal and observe the inflation of the ampullae and tube feet.

Place a small piece of skin on a cavity slide in a drop of water and observe spines and *pedicellariae* under the low power of the microscope, using top illumination.

Place a small piece of skin in a watch-glass containing dilute hydrochloric acid and observe the solution of the chalky parts by the action of the acid.

13 Chordata

Triploblastic coelomate animals, the nervous system formed from a tube of tissue placed dorsally. Below it, a cylindrical rod called the notochord *forms the main axis of an internal skeleton. The coelom does not extend into a tail, placed behind the anus.*

Basic plan of a Vertebrate animal, with a considerable portion of the right-hand side removed to show the internal structure.

The digestive system, spinal cord and vertebrae have been cut near the tail, to show the structure in transverse section.

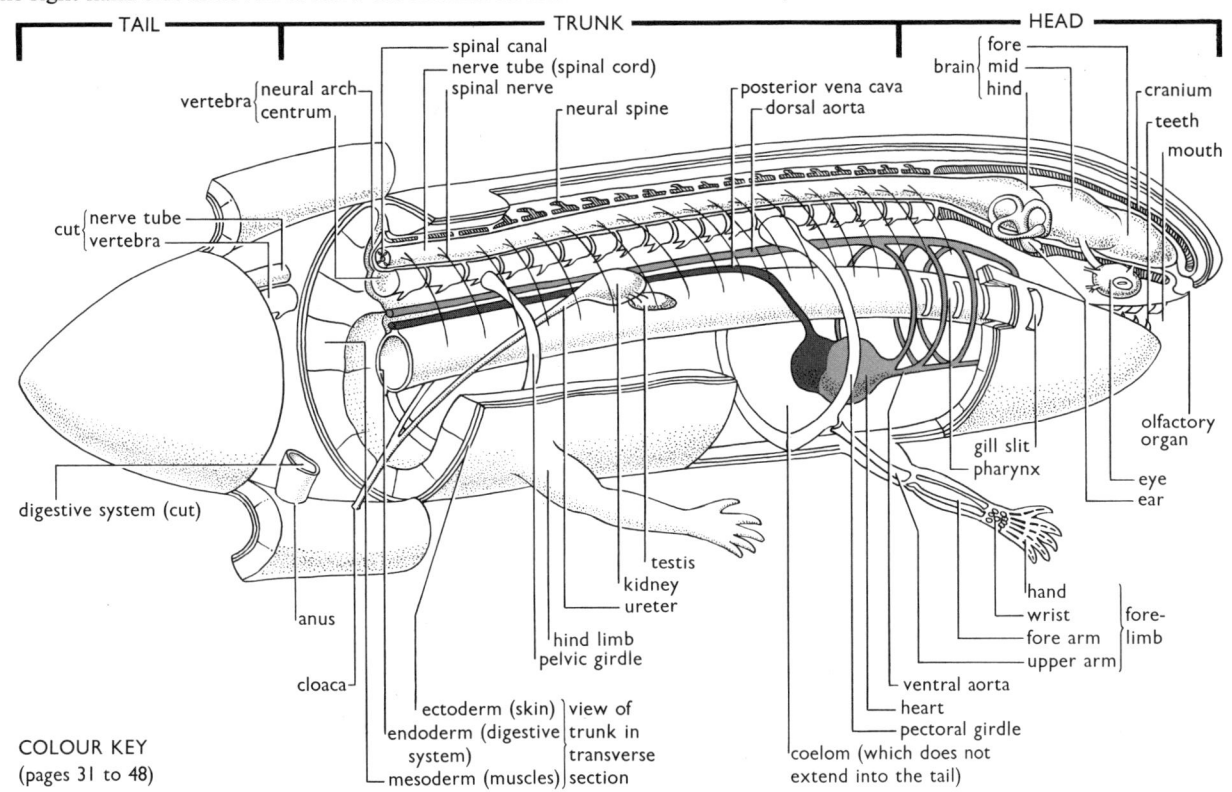

COLOUR KEY
(pages 31 to 48)

▨ Thick walled arteries. (Also, on this page and page 47, the muscular ventricles of the heart.)

▨ Thin walled veins. (Also, on this page and page 47, the thin walled auricles of the heart.)

SUB PHYLUM VERTEBRATA

This Group contains a variety of animals commonly called the *Vertebrates*, because the *notochord* in the adult is divided into a number of articulated segments, the *vertebrae*.

Irritability The dorsal nervous system consists of a hollow tube of nervous tissue, expanded at the anterior end to form a brain. The head carries the principal sense organs in pairs: olfactory organs (smell); eyes (sight); ears (sound and balance). There are also sense organs in the mouth (taste) and on the skin (touch, pain, temperature).

Movement Vertebrates have a jointed internal skeleton, the joints operated by a system of muscles. It is adapted for swimming in the case of fish, flight in birds and walking and crawling in land forms.

Nutrition The digestive system consists of the following main parts:
The mouth surrounded by a pair of jaws which carry teeth in many species. There is a muscular tongue in its ventral region.
The pharynx communicating with the environment by pairs of gill slits in aquatic species. The gill slits are present in the embryos of land forms.
The oesophagus. A tube conveying food from the pharynx.
The stomach. A thick muscular bag, where digestion begins.
The intestine. A tube in which food digestion is completed, and the products absorbed into the body.

The rectum. The terminal part of the digestive system, opening to the exterior at the anus.
There are also two important glands associated with the digestive system, the liver and the pancreas.

Respiration There is a well-developed blood system consisting of a muscular heart placed ventrally in the front part of the body. This circulates blood in a series of *arteries*, vessels which lead away from it at the anterior end of the body, pass upwards round the gill slits and then back along a dorsal artery. The blood then passes through the tissues in fine tubes called *capillaries*, and returns to the heart in *veins*. The blood itself consists of a fluid, *plasma*, in which two kinds of blood cells are suspended: the *white corpuscles* and *red corpuscles*. The latter contain a respiratory pigment, *haemoglobin*, which enables them to carry oxygen round the body. The respiratory organs consist of gills in aquatic species and lungs in land species.

Excretion A pair of kidneys placed dorsally in the body filters liquid excretory materials out of the blood, which are passed to the exterior via ducts called *ureters*.

Reproduction The sexes are separate. The male produces sperms in a pair of testes, and these pass out of the body through sperm ducts, which are closely connected to the *ureters* in some species. The female produces eggs in a pair of ovaries, the eggs pass out of the body through oviducts, each one with its internal opening near the ovary.

13.1 CLASS OSTEICHTHYES (Bony Fish)

Aquatic Vertebrates which breathe by means of gills.

13.1.1 The Herring (*Clupea harengus*)

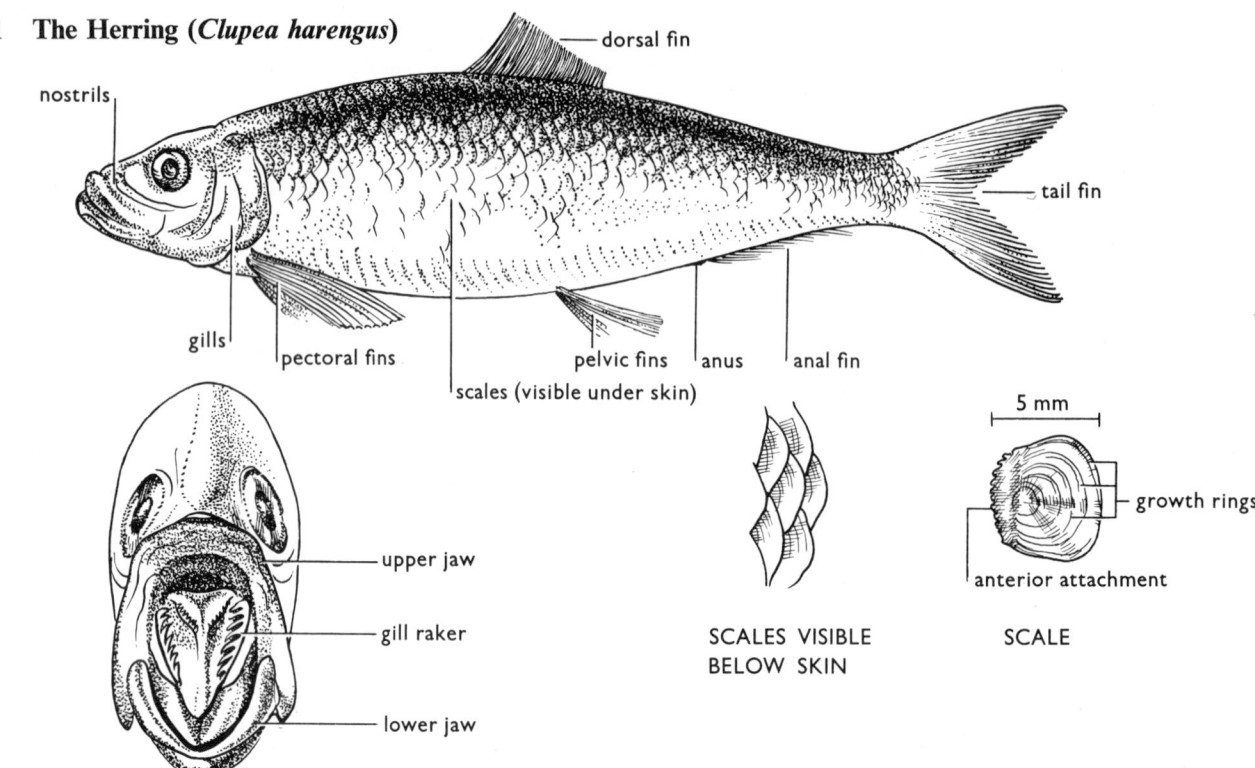

SCALES VISIBLE BELOW SKIN

SCALE

ANTERIOR VIEW WITH MOUTH WIDE OPEN

A marine fish which lives in shoals consisting of millions of individuals. The body is covered with a smooth slimy skin, beneath it lie fine semi-circular bony scales marked with rings, each ring representing one year in the life of the fish. It swims by means of fins, the most important being at the tip of its powerful muscular tail.

It detects vibrations in the water by means of sense organs in a pair of lateral lines, one down each side of the body. It feeds on plankton, small plants and animals near the surface of the sea, which it strains from the current of water passing over its gills by means of gill rakers. There are four pairs of gills, delicate structures protected by a flap of skin on each side of the head, called the *operculum*. The females lay eggs in vast quantities at the bottom of the sea, and, when they hatch out, the larvae swim to the surface to feed on plankton.

Examine the external features of a herring, noting the structure of the mouth, gills and fins. With a pair of forceps make an incision in the skin at one side of the body. Remove a number of scales from beneath the skin with the forceps and mount them dry on a microscope slide. Examine the growth rings either with a hand lens or under the low power of the microscope.

13.1.2 The Minnow (*Phoxinus laevis*)

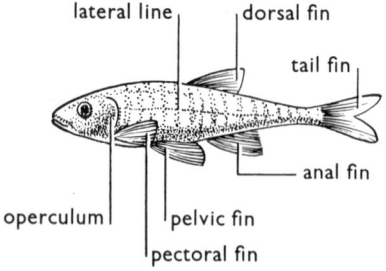

A freshwater fish which is very common in ponds and streams and reaches a length of approximately 8 cm It lives in small shoals and moves actively in search of food, feeding on small water animals and plants. It breathes by means of gills but it is also able to gulp air in through its mouth, which is swallowed into the intestine. The walls of the intestine are lined with blood vessels which absorb oxygen from the air, and so enable the fish to live in poorly aerated water. The female lays about 1,000 eggs at a time, usually amongst gravel in shallow water.

Examine living minnows in small glass dishes. During the breeding season, males can be distinguished from females by red colour at the corners of their mouths. Place one dish on a white background and another on a dark one and observe the change in colour of these animals. Also observe colour change when the fish are fed with morsels of food. Examine the external features, noting the structure and arrangement of the fins.

13.2 CLASS AMPHIBIA (Frogs and Newts)

Vertebrates which spend their larval life in fresh water, but the adults are able to live out of water.

13.2.1 The Common Frog (*Rana temporaria*)

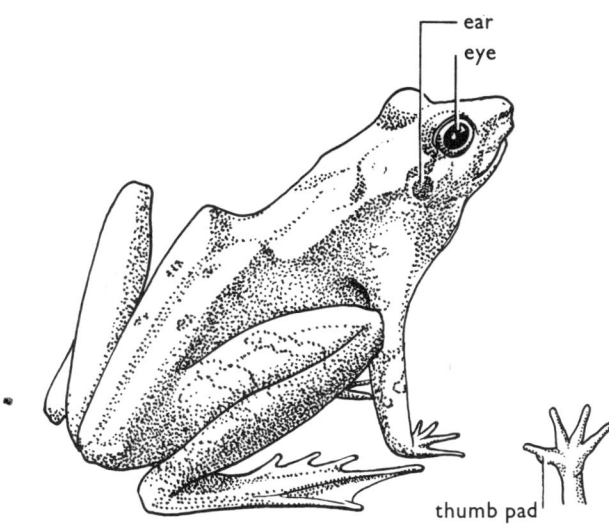

MALE FROG SEEN FROM RIGHT SIDE   LEFT HAND

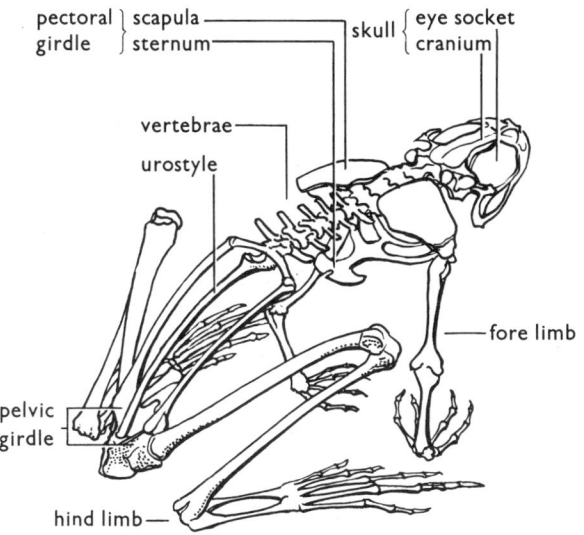

SKELETON, SEEN FROM RIGHT SIDE

Habitat The larva, or tadpole, lives in ponds and streams, the adult lives in and around damp places.

External appearance The adult has a squat body covered by a damp, slimy skin. There is no tail. The hind limbs have webbed feet with 5 digits. The fore limbs have 4 digits, the male can be identified by pads on its thumbs.

Skeleton There is an internal bony skeleton. The skull is broad, with a centrally placed cranium with a large eye-socket on each side of it. Behind it and running along the dorsal axis of the animal is a vertebral column consisting of 9 small vertebrae, with a long *urostyle* at the posterior end. The fore limbs are comparatively short and are attached to the pectoral girdle. The hind limbs are larger and are attached to the pelvic girdle, rigidly fixed to the posterior end of the vertebral column.

CENTRAL NERVOUS SYSTEM, DORSAL ASPECT

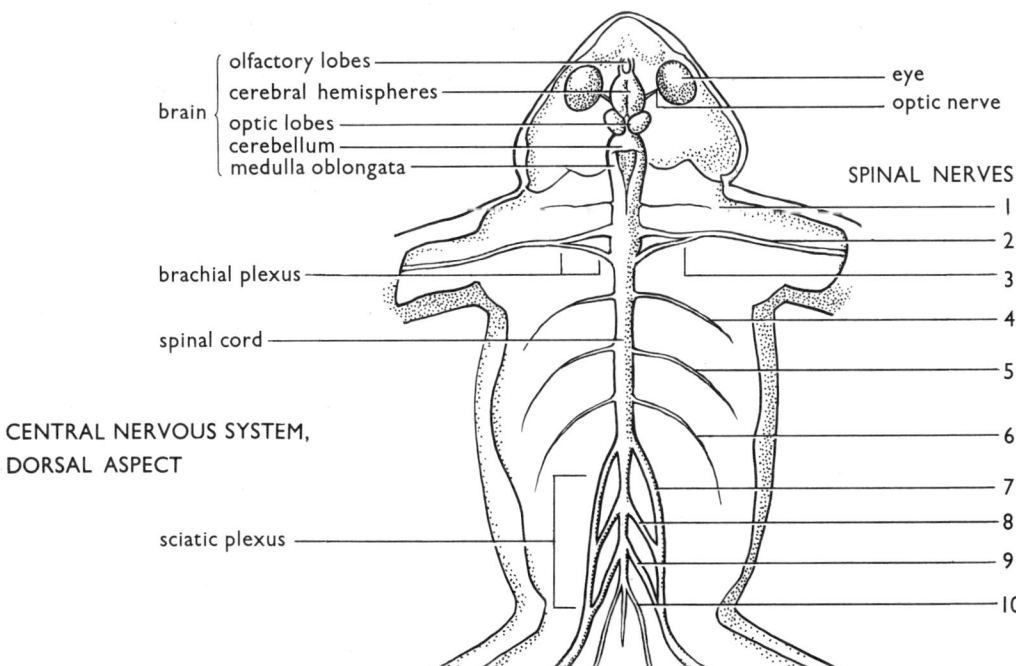

CENTRAL NERVOUS SYSTEM, DORSAL ASPECT

Irritability There is a prominent pair of eyes. Below each one is an ear-drum. The central nervous system is well developed, with a brain lying inside the cranium. Ten pairs of nerves arise from the spinal cord. Examine a prepared dissection of the nervous system, made from the dorsal surface of the animal.

33

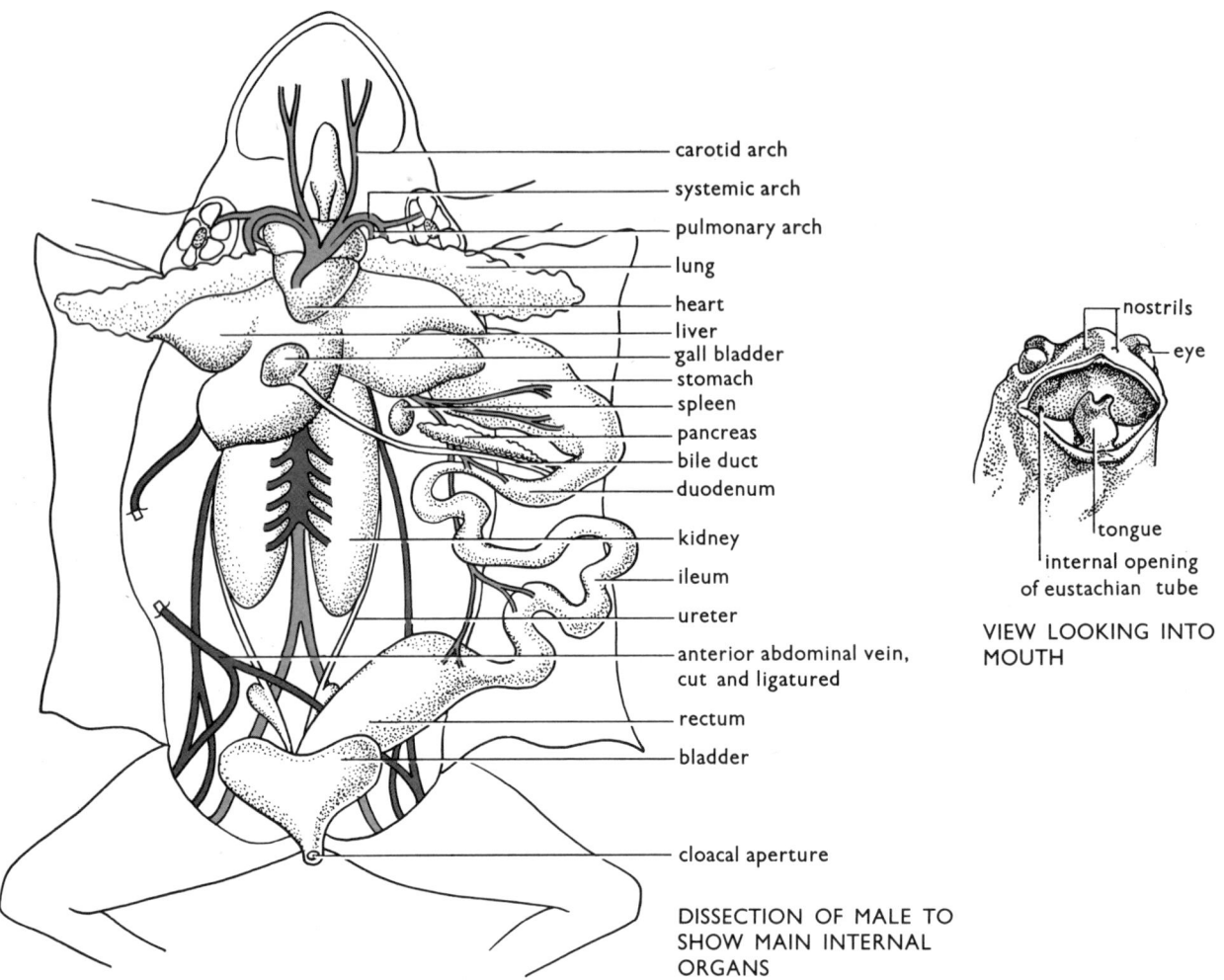

- carotid arch
- systemic arch
- pulmonary arch
- lung
- heart
- liver
- gall bladder
- stomach
- spleen
- pancreas
- bile duct
- duodenum
- kidney
- ileum
- ureter
- anterior abdominal vein, cut and ligatured
- rectum
- bladder
- cloacal aperture

DISSECTION OF MALE TO SHOW MAIN INTERNAL ORGANS

- nostrils
- eye
- tongue
- internal opening of eustachian tube

VIEW LOOKING INTO MOUTH

Nutrition Its main food is flies, which are caught while they are on the wing with a quick action of its sticky tongue, which is hinged near the *front edge* of the lower jaw.

Method of dissection Place a freshly killed frog dorsal surface downwards in a dissecting dish. Pin it to the wax with a pin through each limb, so that the whole animal is extended. Cover with water containing a pinch of salt. Open the mouth with a pair of forceps, and examine the structure of the buccal cavity.

To start the dissection, lift a piece of skin with the forceps in the middle of the abdominal region. Snip into it with a pair of scissors, and make a cut along the mid-line for the whole length of the body. Make a lateral cut part way along each forearm and for a short distance down each leg. Be particularly careful not to dig your scissors downwards, especially in the region of the armpits, where important blood vessels pass into the skin at this point. Pin the skin back down each side of the body.

Now dissect open the body wall. To do this, carefully make two cuts about 1 cm long, one on each side of the anterior abdominal vein, which will be clearly visible running along the axis of the body wall from front to rear. With a pair of forceps, pull a loop of thread through the cut on one side, under the vein and out through the cut on the other side. Cut the loop of thread, so that you now have two pieces of thread passing under the vein. Tie a knot in each piece of thread so as to constrict the vein. Now cut the vein through with the scissors, and, if the threads have been properly tied, there will be no leakage of blood. Cut forwards along one side of the vein, until you reach the sternum, when you can safely continue your cut in the mid-line of the body and so through the ventral part of the pectoral girdle. Undo the pins holding the arms and pull the arms apart gently and pin them down again. Pin the body wall out to each side of the animal, so as to reveal the internal organs. With the aid of a seeker and a hand lens examine their general arrangement and identify them.

If the water in your dissecting dish becomes cloudy during the dissection, change it for fresh water and put a pinch of salt in it, as before.

Respiration

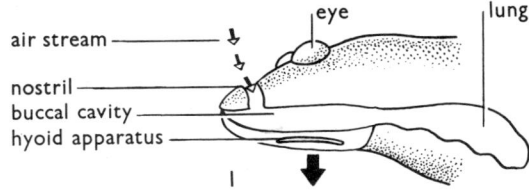

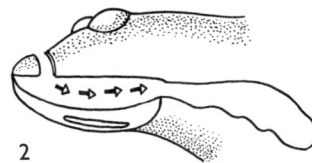

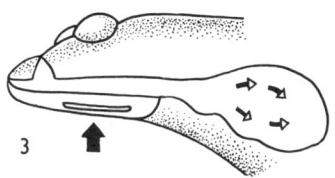

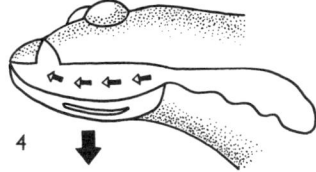

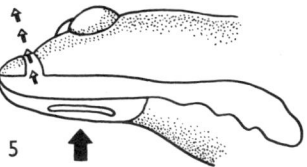

The frog uses three methods for breathing: (a) through its damp skin; (b) through the lining of the buccal cavity; (c) through the tissues of the lungs. Observe a living frog and notice the movements of the nostrils and hyoid apparatus of the lower jaw when it is breathing through its lungs. The diagrams above represent the cycle of events during lung breathing.

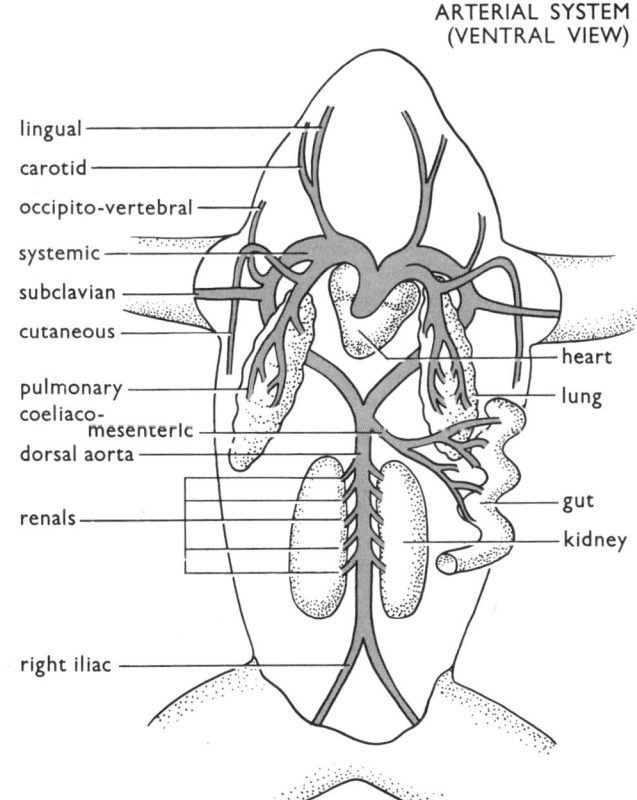

ARTERIAL SYSTEM
(VENTRAL VIEW)

lingual
carotid
occipito-vertebral
systemic
subclavian
cutaneous
pulmonary
coeliaco-mesenteric
dorsal aorta
renals
right iliac

heart
lung
gut
kidney

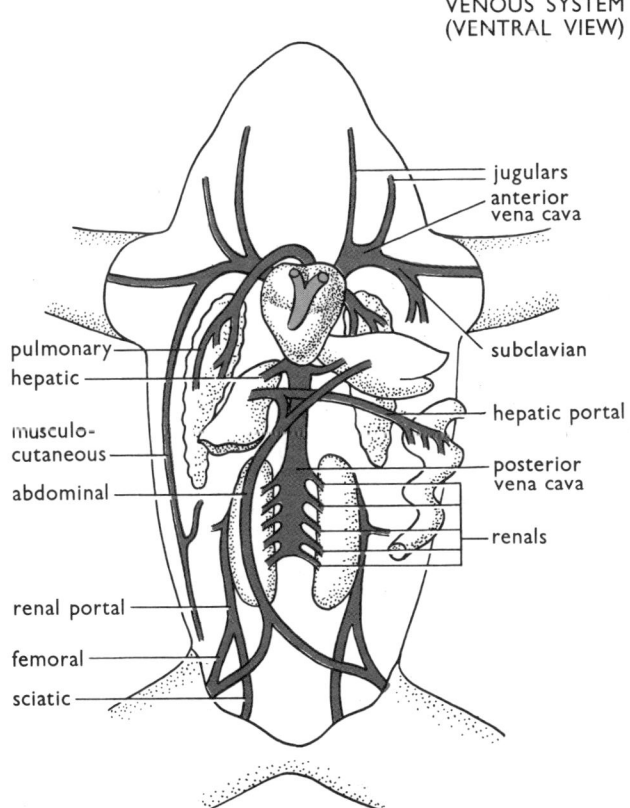

VENOUS SYSTEM
(VENTRAL VIEW)

jugulars
anterior vena cava
subclavian
hepatic portal
posterior vena cava
renals

pulmonary
hepatic
musculo-cutaneous
abdominal
renal portal
femoral
sciatic

The respiratory gases are carried round the body in the blood system. The main blood vessels are shown in the diagrams above. Try to identify them in your general dissection. Blood travels *away from* the heart in the *arteries* of the arterial system. It *returns to* it in the *veins* of the venous system.

35

The heart

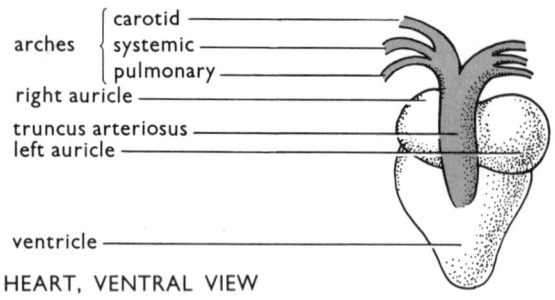

arches { carotid / systemic / pulmonary
right auricle
truncus arteriosus
left auricle

ventricle

HEART, VENTRAL VIEW

The heart lies immediately beneath the sternum, and is revealed when the arms are pulled apart during the later stages of the dissection. (See section on General Dissection.) It consists of 5 parts: 2 *auricles*; a *ventricle*; a *truncus arteriosus*; and a *sinus venosus*—all totally enclosed in a transparent piece of skin called the *pericardium*. With great care this can be dissected away with scissors and forceps. The heart frequently starts beating again, especially if it is massaged by rhythmically pinching the ventricle with the forceps. Blood is pumped by the ventricle into the arteries, in which it travels to all parts of the body and lungs. Blood returning from the body in veins enters the right auricle; that returning from the lungs enters the left auricle.

Frog's blood

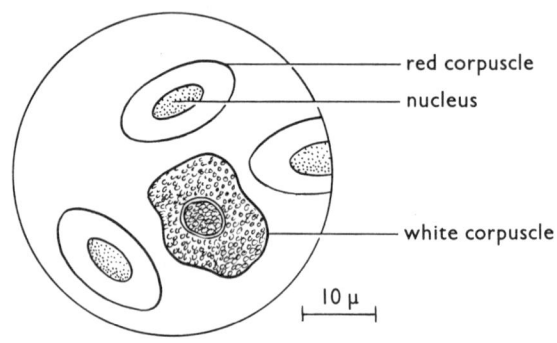

red corpuscle
nucleus

white corpuscle

10 μ

Make a thin smear of frog's blood on a slide. Observe the disc-shaped red corpuscles containing prominent nuclei. Examine the smear carefully for irregular-shaped white corpuscles, which occur less frequently than the red ones. Carry out your observations under the high power of the microscope.

Capillaries in tadpole's tail

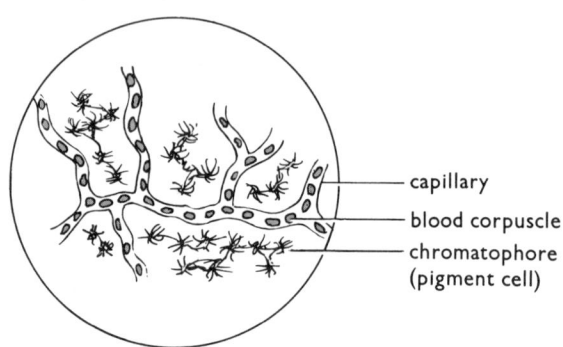

capillary
blood corpuscle
chromatophore (pigment cell)

The arteries are connected to the veins to complete the circuit by means of microscopic tubes called *capillaries*. They lie in most of the tissues of the body. Mount a tadpole in a drop of water in a cavity slide, covering its tail with a cover-slip. Observe the rhythmic flow of the blood in the capillaries, under the low power of the microscope.

Reproduction and excretion

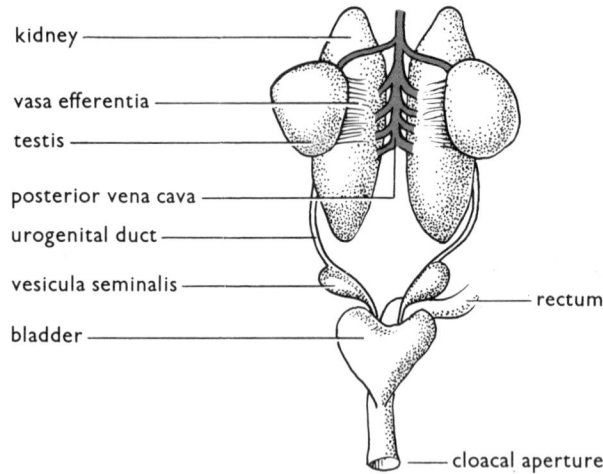

kidney
vasa efferentia
testis
posterior vena cava
urogenital duct
vesicula seminalis
bladder

rectum

cloacal aperture

MALE UROGENITAL SYSTEM
(VENTRAL VIEW)

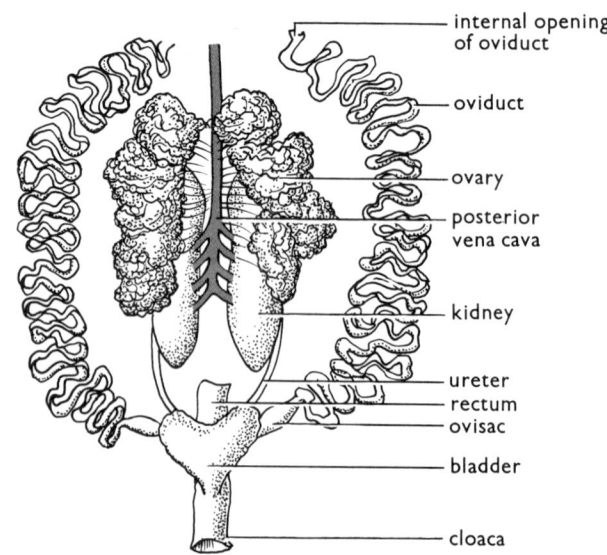

internal opening of oviduct
oviduct
ovary
posterior vena cava
kidney
ureter
rectum
ovisac
bladder
cloaca

FEMALE UROGENITAL SYSTEM
(VENTRAL VIEW)

The sexes are separate. Remove the digestive system by snipping through the rectum at one end and the stomach at the other. Identify the organs of the reproductive system and excretory systems.

Life-cycle

Frogs emerge from hibernation in early spring. The male sits astride the female, gripping her slimy skin by means of the pads on his hands. After a time, the female lays about 1,000 eggs which pass out of the *cloaca* into the pond water. At the same time, the male sheds sperms from his cloaca, and these swim in the water and reach and fertilize most of the eggs. The jelly-like spawn swells, helping to make the eggs buoyant.

Observe the development of the eggs in a small jar or aquarium; the following is an approximate timetable.

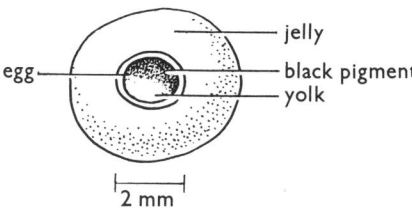

Day 1   Yellow yolk visible in the lower third of the egg.

Day 2   Yolk now a small plug at the equator of the egg.

Day 7   Embryo elongating; head, body and tail visible.

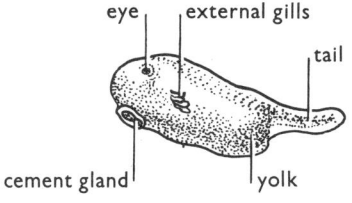

Day 14   The tadpole hatches by wriggling through the spawn. The *cement gland* attaches it to the surface of the spawn. The mouth is not yet open, so it still feeds on *yolk* inside the body. There are three pairs of *external gills*. If these are examined under the low power, blood can be seen circulating in their capillaries.

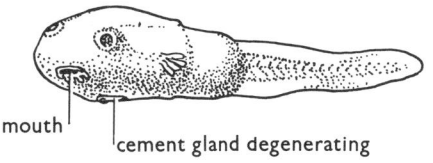

Day 18   The mouth opens and the tadpole begins to swim about, feeding on vegetable matter.

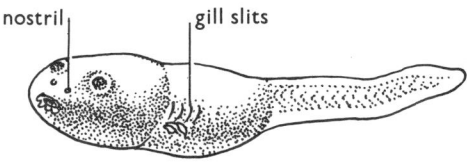

Day 22   The external gills are gradually replaced by four pairs of *gill slits*, which develop on each side of the body.

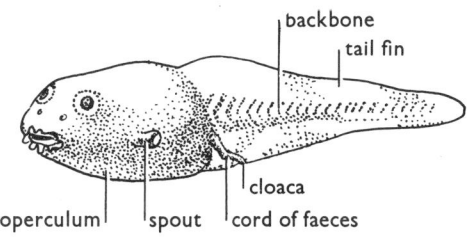

Day 28   A fold of skin called the *operculum* grows over the front part of the body, enclosing and protecting the gill slits. In breathing, water is drawn in through the mouth, over the gill slits, and out through a spout on the left-hand side.

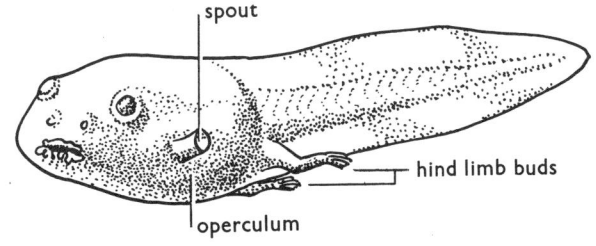

Day 50   A pair of *hind limb buds* are visible, the fore-limbs develop underneath the operculum and are not visible.

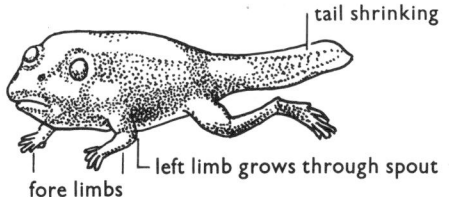

Day 90   Metamorphosis. The tadpole rapidly adapts itself to the adult terrestrial form. The fore-limbs grow through the operculum, the tail shrinks and is converted into adult body material. The eyes and mouth become prominent. The young adult feeds on insects and, after 3 years' development, is able to reproduce.

13.3 CLASS REPTILIA (Reptiles)

Terrestrial Vertebrates, with teeth in their jaws and the body covered with dry scales.

13.3.1 The Common Lizard (*Lacerta vivipara*)

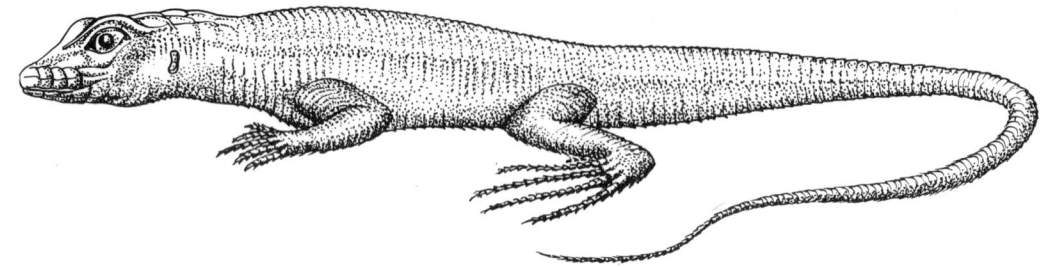

It reaches a length of 15 cm , and in warm weather is active amongst vegetation. It has prominent eyes protected by eyelids, a sensitive forked tongue and a pair of ears. It runs actively in search of insects, its main food. Air is breathed into the body through a pair of lungs. Eggs are fertilized in the female's body by sperms injected by the male. The egg develops inside a calcareous egg shell, the embryo hatches through it, inside the mother's body.

The Common Lizard is *viviparous*, i.e. the young are born alive (usually 6 to 12), together with the broken egg shells.

Examine a live or preserved specimen. Note the eyes, nostrils, teeth, tongue and ears. Note the nature of the scales, limbs and tail. The tail has a breaking point near its base, which often snaps if it is caught by a bird. A new tail, though poor in appearance, sometimes regenerates from the stump.

13.4 CLASS AVES (Birds)

Vertebrates having the forelimbs modified as wings. The body covered with feathers, its temperature controlled at approximately 40°C.

13.4.1 The Common Pigeon (*Columba livea*)

cranium

orbit

jaws

cervical vertebrae

thoracic vertebra

rib

fore limb

sacrum

sternum

pelvic girdle

femur

tibio-tarsus

ankle joint

foot

pygidium

caudal vertebrae

SKELETON, SEEN FROM RIGHT SIDE

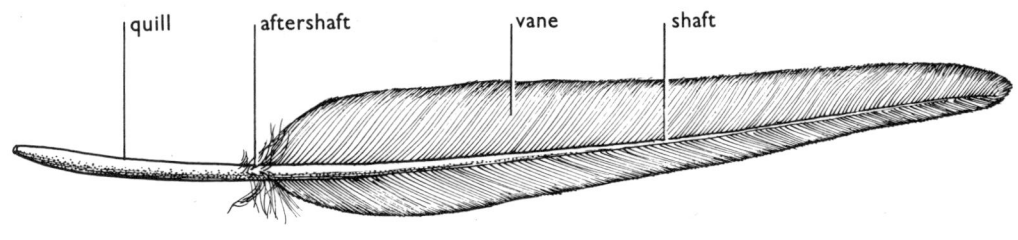

A WING FEATHER

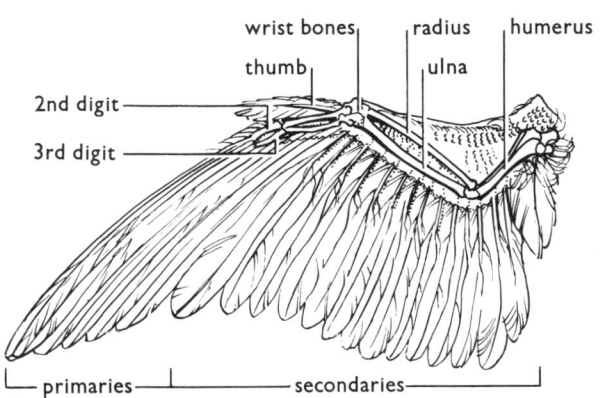

WING, TO SHOW FEATHER ARRANGEMENT

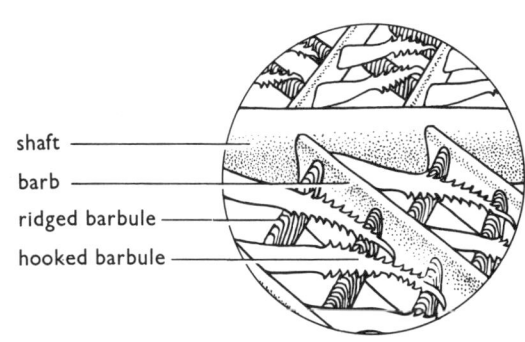

PORTION OF FEATHER MAGNIFIED

Habitat and structure It is a domesticated form of the natural Rock Pigeon. The body is light and streamlined, and is covered with feathers which keep it warm as well as being used in flight. A typical wing feather consists of a flat *vane* supported by a solid *shaft* continued into a hollow tube called the *quill*. The quill attaches the feather to the skin in a small pit called the *follicle*. The vane is made up of slanting pieces called *barbs*, carrying rows of smaller *barbules*. Those facing the tip of the feather have hooks which fit into ridges on the barbules of the adjacent barbs, so locking together to make a continuous vane. If the barbs are separated, they can be rejoined by stroking them as in preening, an action which the bird performs with its beak.

Irritability and movement Each eye has an upper and lower eyelid and also a thin transparent *nictitating membrane*, which can be extended from the innermost corners and is used for cleaning the cornea. The ear openings are placed behind the eyes. The sense of smell is poorly developed. The wings are very efficient organs of flight operated by powerful muscles attached to the pectoral girdle.

Nutrition It feeds on grain, which is stored in a large *crop* at the base of the neck. Here it is partially digested and passed to the stomach. There is a horny beak surrounding the mouth, inside which is a pointed tongue. There are no teeth.

Respiration Air is drawn into the lungs by movements of the sternum. The lungs have extensions into the body tissues called *air-sacs*. There is a *syrinx* (voice box) at the base of the wind-pipe.

Excretion and reproduction There is a pair of large kidneys, each joined to the *cloaca* by a *ureter*. There is no bladder. The eggs are fertilized inside the female's body. There is one large ovary and oviduct, the sex organs of the right side do not develop. Eggs are laid two at a time in a rough nest and incubated by the mother at about 40°C. They hatch out into young pigeons covered with a fine down (feathers without vanes), in about 18 days.

Method of examination Examine a live or preserved pigeon, noting the external features. Study the structure of a feather with a hand lens. Study the interlocking of the barbules, by cutting a small portion of a vane and mounting it dry on a slide. Observe it through the low power of the microscope. Examine a preserved skeleton of a pigeon or similar bird.

39

13.5 CLASS MAMMALIA (Mammals)

Vertebrates whose bodies are covered with hair. The temperature kept constant at about 35°C. The young develop inside the reproductive system of the female and are born in an advanced state. They are then fed on milk produced in mammary glands on the ventral surface of the female body. Lungs and heart lie in a thorax separated from the abdomen by a diaphragm.

NOTE: The Rat is taken as the main example from this class, but the Rabbit is more suitable for the study of the skeleton and the Sheep for the study of eye and heart. This is noted in the section concerned.

13.5.1 The Rat (*Rattus norvegicus*)

Habitat It lives in and around human dwellings, where it feeds on food scraps and rubbish. The domesticated albino variety is docile and is the best for study alive, and for dissection.

The skeleton The main functions of the Mammalian Skeleton are: (*a*) to support the tissues of the body; (*b*) to allow movement through a system of joints; (*c*) to provide anchorage for muscles; (*d*) to protect vital organs.

The bones which form the various parts of a skeleton are made from:

(1) *Organic materials*, giving flexibility; they are insoluble in dilute acids.
(2) *Inorganic materials* (calcium salts), which give rigidity, and are soluble in dilute acids.

Place a small bone in a beaker of dilute acid, and note its flexibility after the acid has dissolved the inorganic materials. Place a small bone in a crucible and heat it strongly to destroy the organic material. Note the inorganic ash which remains. (For microscopic structure, see page 52.)

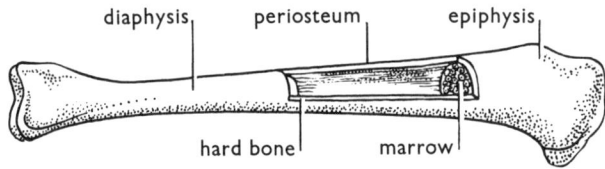

DIAGRAM TO SHOW STRUCTURE OF LONG BONE

Joints Identify the tissues surrounding a joint, in the hind limb of a rat, during general dissection.

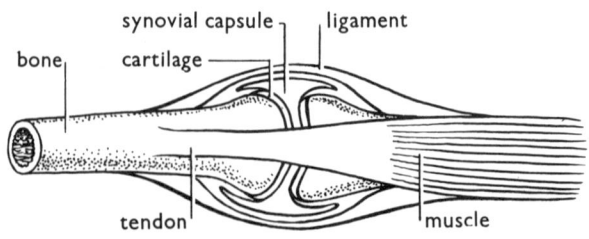

DIAGRAM TO SHOW STRUCTURE OF A JOINT

40

The Skeleton of the Rabbit (*Oryctolagus cuniculus*)

The skeleton consists of two sections:
(1) The Axial Skeleton,
(2) The Appendicular Skeleton.

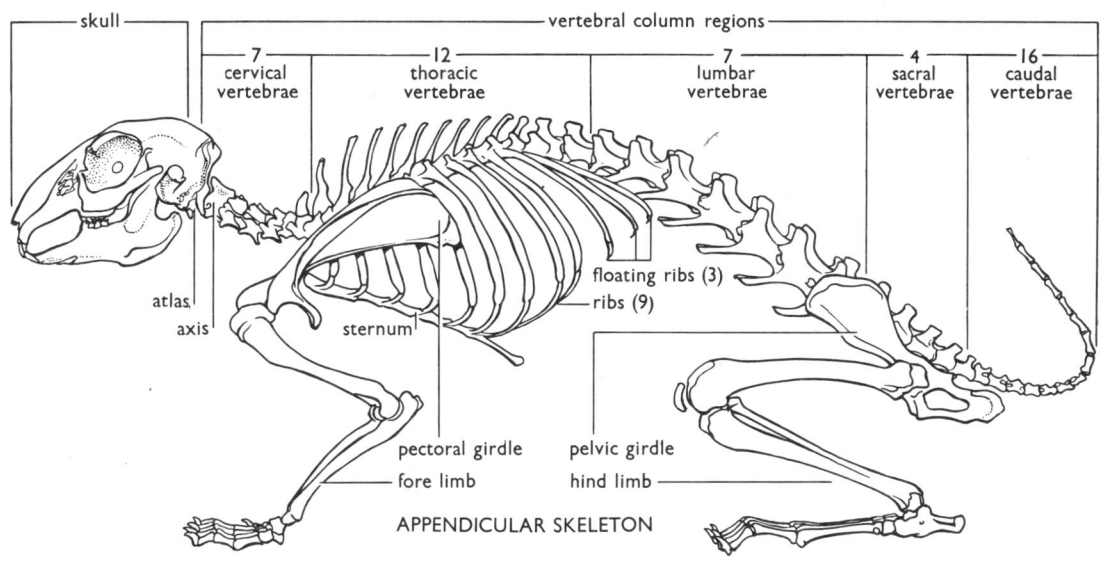

RABBIT SKELETON TO SHOW AXIAL SKELETON
AND LEFT SIDE OF APPENDICULAR SKELETON

The axial skeleton   This occupies the central axis of the animal, and protects the *Central Nervous System* (see page 43), the organs of the thorax, as well as providing the main support for the limbs and limb girdles.

(a) The skull

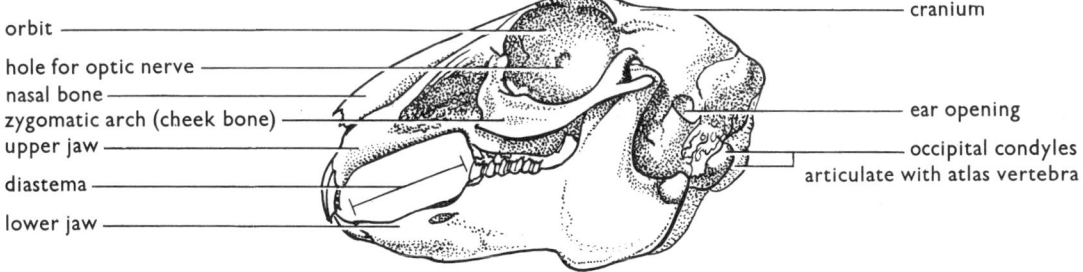

RABBIT SKULL VIEWED FROM LEFT SIDE

(b) The vertebral column

This consists of a number of bony segments called *vertebrae*. A typical vertebra has a hard *centrum* on its ventral side, giving rigidity to the vertebral column as a whole. On the dorsal side an arch of bone (the *neural arch*) surrounds and protects the *spinal cord*. Between each vertebra, a biconvex lens-shaped piece of cartilage lies between each centrum. The neural arches articulate with each other at surfaces called *zygapophyses*.

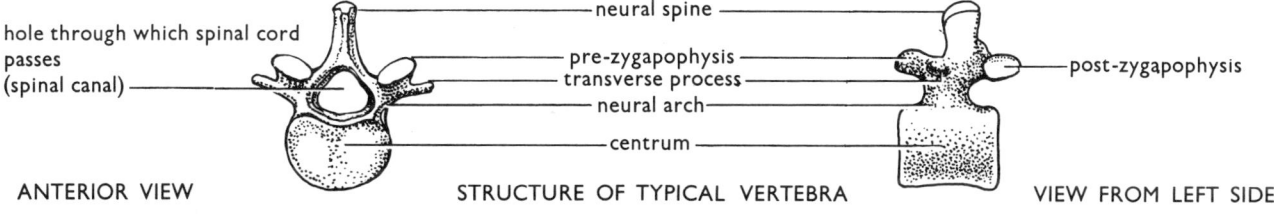

ANTERIOR VIEW   STRUCTURE OF TYPICAL VERTEBRA   VIEW FROM LEFT SIDE

41

The vertebral column [continued]

It is divided into five regions:

Cervical (neck): 7 vertebrae.   Sacral: 4 vertebrae fused together.
Thoracic (chest): 12 or 13 vertebrae.   Caudal (tail): 16 vertebrae approximately.
Lumbar (back): 6 or 7 vertebrae.

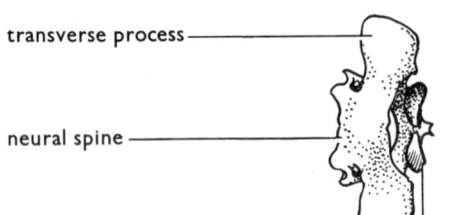

transverse process

neural spine

post zygapophysis

1st CERVICAL VERTEBRA
(ATLAS) DORSAL VIEW

surface which articulates with the atlas

cervical rib

neural spine

odontoid peg

2nd CERVICAL VERTEBRA
(AXIS) DORSAL VIEW

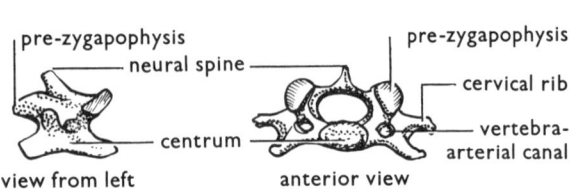

pre-zygapophysis
neural spine

pre-zygapophysis

cervical rib

centrum

vertebra-
arterial canal

view from left   anterior view

5th CERVICAL VERTEBRA

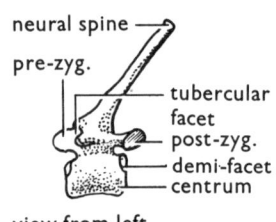

neural spine
pre-zyg.

tubercular
facet
post-zyg.
demi-facet
centrum

view from left

6th THORACIC VERTEBRA

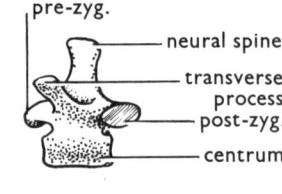

pre-zyg.

neural spine

transverse
process
post-zyg.

centrum

view from left

12th THORACIC VERTEBRA

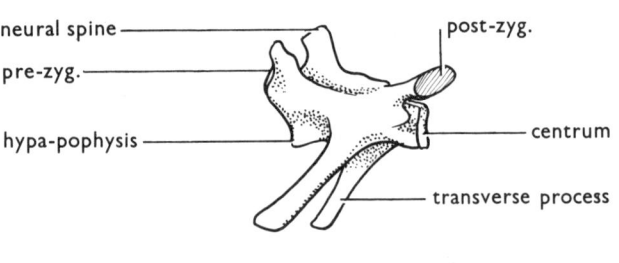

neural spine
pre-zyg.

post-zyg.

hypa-pophysis

centrum

transverse process

2nd LUMBAR VERTEBRA   view from left

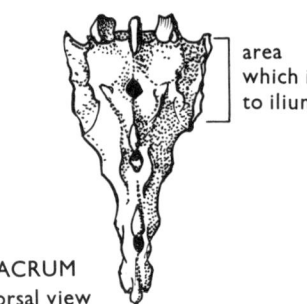

area
which is fused
to ilium

SACRUM
dorsal view

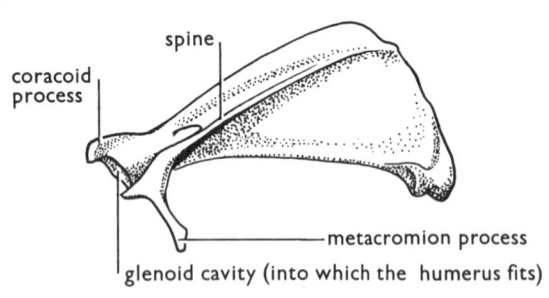

3 CAUDAL VERTEBRAE

The appendicular skeleton

This consists of the pectoral girdle and fore-limbs, the pelvic girdle and hind-limbs.

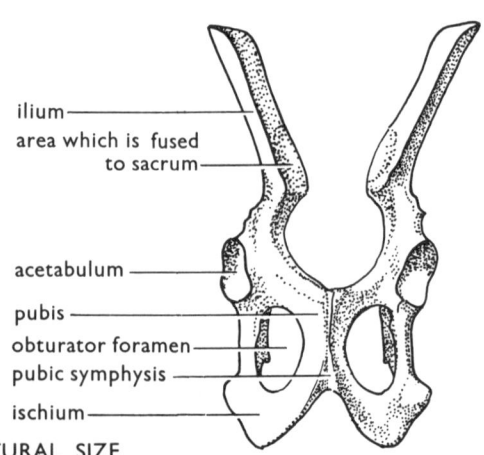

spine
coracoid
process

metacromion process
glenoid cavity (into which the humerus fits)

ilium
area which is fused
to sacrum

acetabulum

pubis
obturator foramen
pubic symphysis
ischium

ALL DRAWINGS NATURAL SIZE

PECTORAL GIRDLE, LEFT SCAPULA   PELVIC GIRDLE, VENTRAL VIEW

The appendicular skeleton [*continued*]

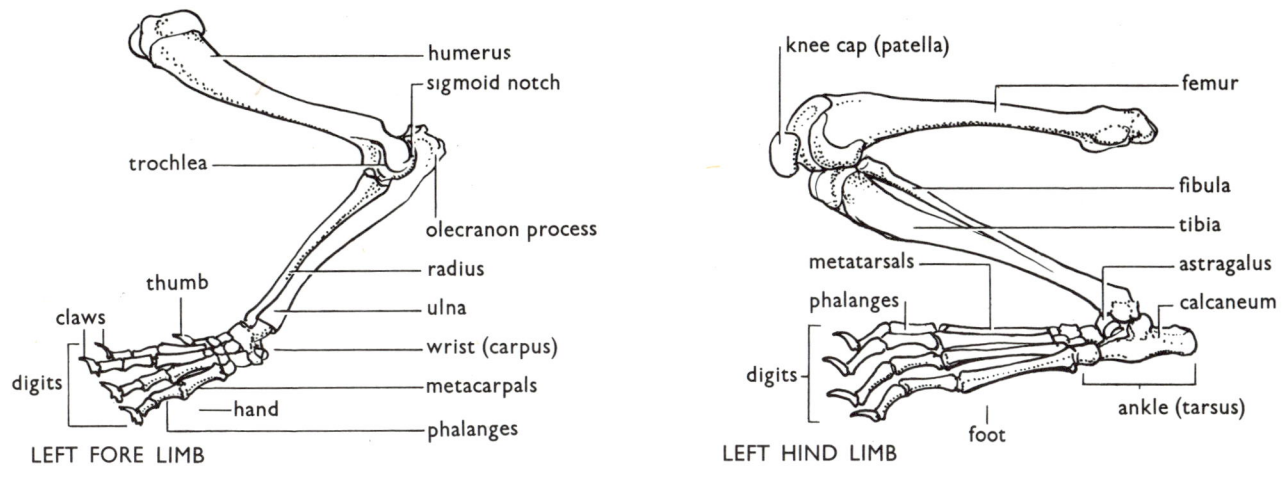

LEFT FORE LIMB

LEFT HIND LIMB

Irritability The nervous system consists of three main parts: the *Central* Nervous System; the *Peripheral* Nervous System; and the *Autonomic* Nervous System. The layout of these three systems is shown in the diagram below.

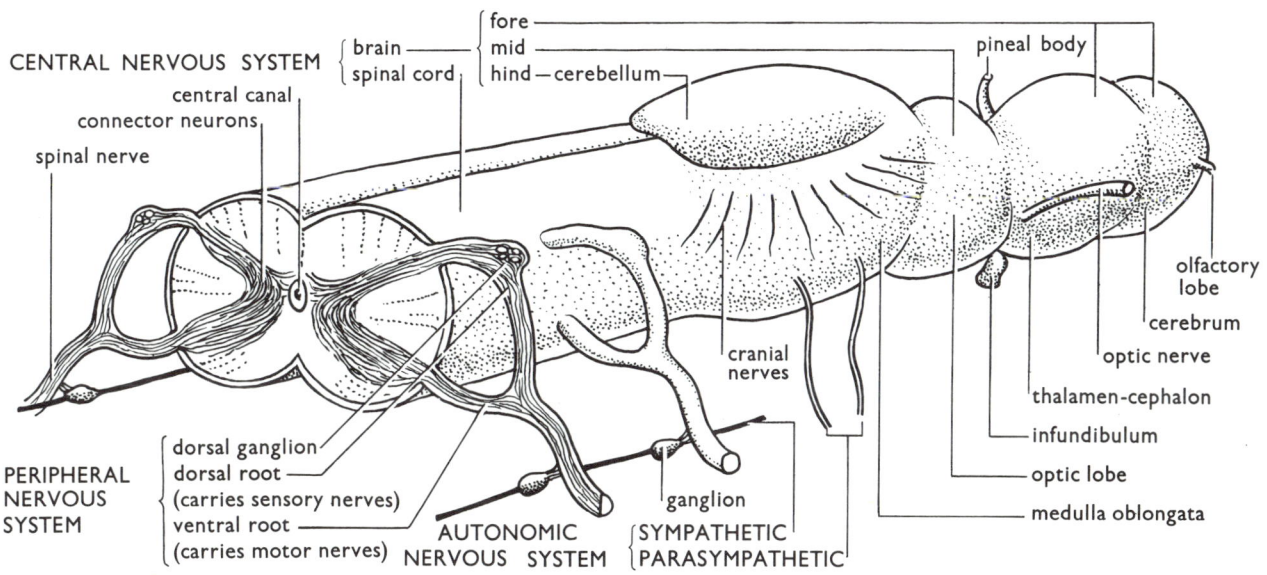

The Central Nervous System is made from a hollow tube of nervous tissue, the *spinal cord*. It runs along the dorsal axis. At the anterior end it expands to form the *brain*, consisting of three regions: fore-, mid- and hind-brain. The cranial nerves extend from these to the sense organs and muscles in the head, and viscera in the body.

The Peripheral Nervous System consists of nerves which radiate throughout the body, those entering the dorsal part of the spinal column come from sense organs, those leaving the

ventral part go to voluntary muscles. (See page 53.) The dorsal and ventral nerve roots enter the spinal column between vertebrae.

The Autonomic Nervous System is an independent system concerned mainly with supplying involuntary muscles. (See page 53.) It consists of two systems, the *sympathetic* and *parasympathetic*. Examine and study the preserved central nervous system of a rabbit or rat.

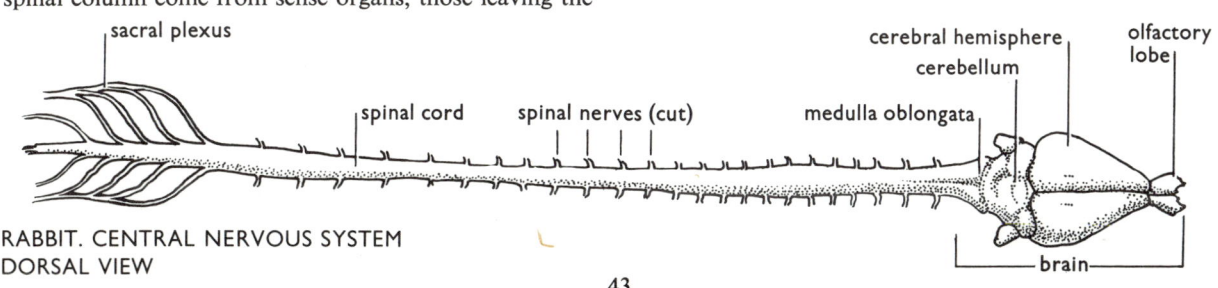

RABBIT. CENTRAL NERVOUS SYSTEM
DORSAL VIEW

The Mammalian Eye

SHEEP'S EYE

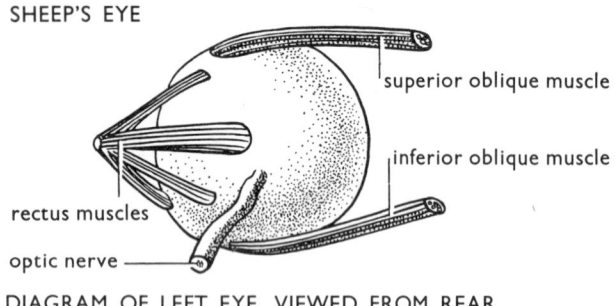

DIAGRAM OF LEFT EYE VIEWED FROM REAR

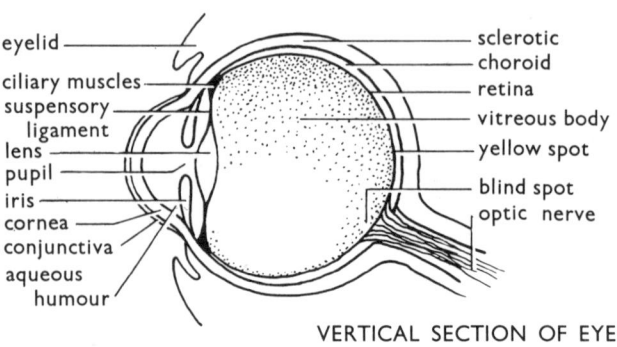

VERTICAL SECTION OF EYE

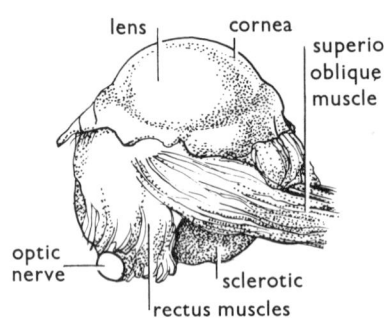

VIEW OF WHOLE EYE

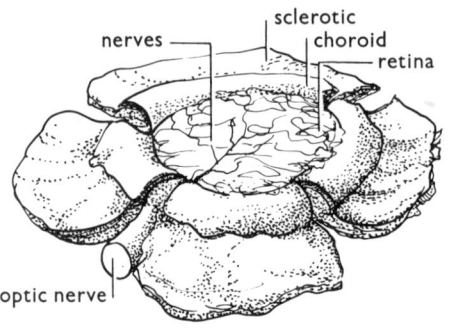

DISSECTION OF EYE

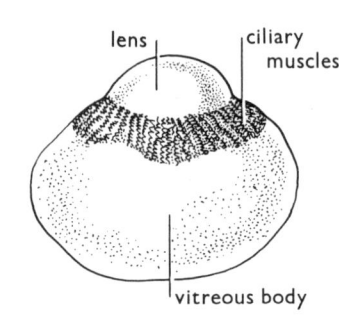

VITREOUS BODY AND LENS REMOVED

The sheep's eye is a convenient type to dissect. Examine a fresh eye, and identify the *rectus* and *oblique muscles*, and the *optic nerve*. Dissect the eye in a dissecting dish, by making a cut with a pair of scissors round the circumference of the *cornea* and remove it to expose the *lens* and brown *aqueous humour*. Make four cuts equidistant from each other round the surface of the *sclerotic* towards the optic nerve. Fold the four sections back and pin them to the wax. Remove the *vitreous body* which will often have the lens and *ciliary muscles* adhering to it. Identify the *retina* (green in colour), the *choroid* (black) and the *blind spot*, together with the nerves and blood vessels entering the eyeball at this point. Examine a prepared slide (see page 54) under the microscope.

The Mammalian Ear

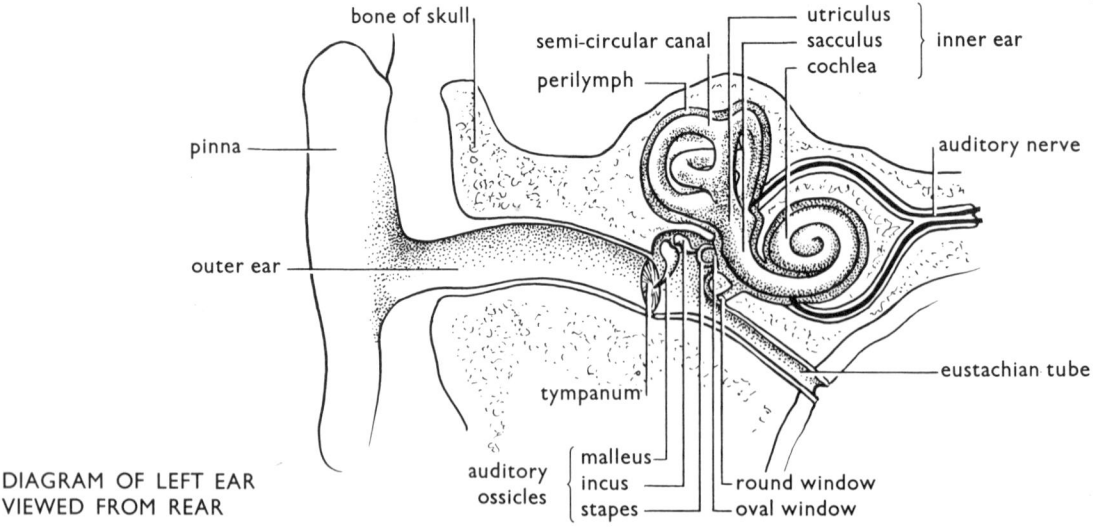

DIAGRAM OF LEFT EAR
VIEWED FROM REAR

Study the structure of the pinna and outer ear in a preserved rat. Carefully insert a pair of forceps into the middle ear, and remove the ear *ossicles* (malleus, incus, stapes). Mount them in a drop of water on a slide and examine under the low power of the microscope. Examine prepared slides of the *cochlea* (see page 54).

44

Method of Examination of a Rat by a General Dissection

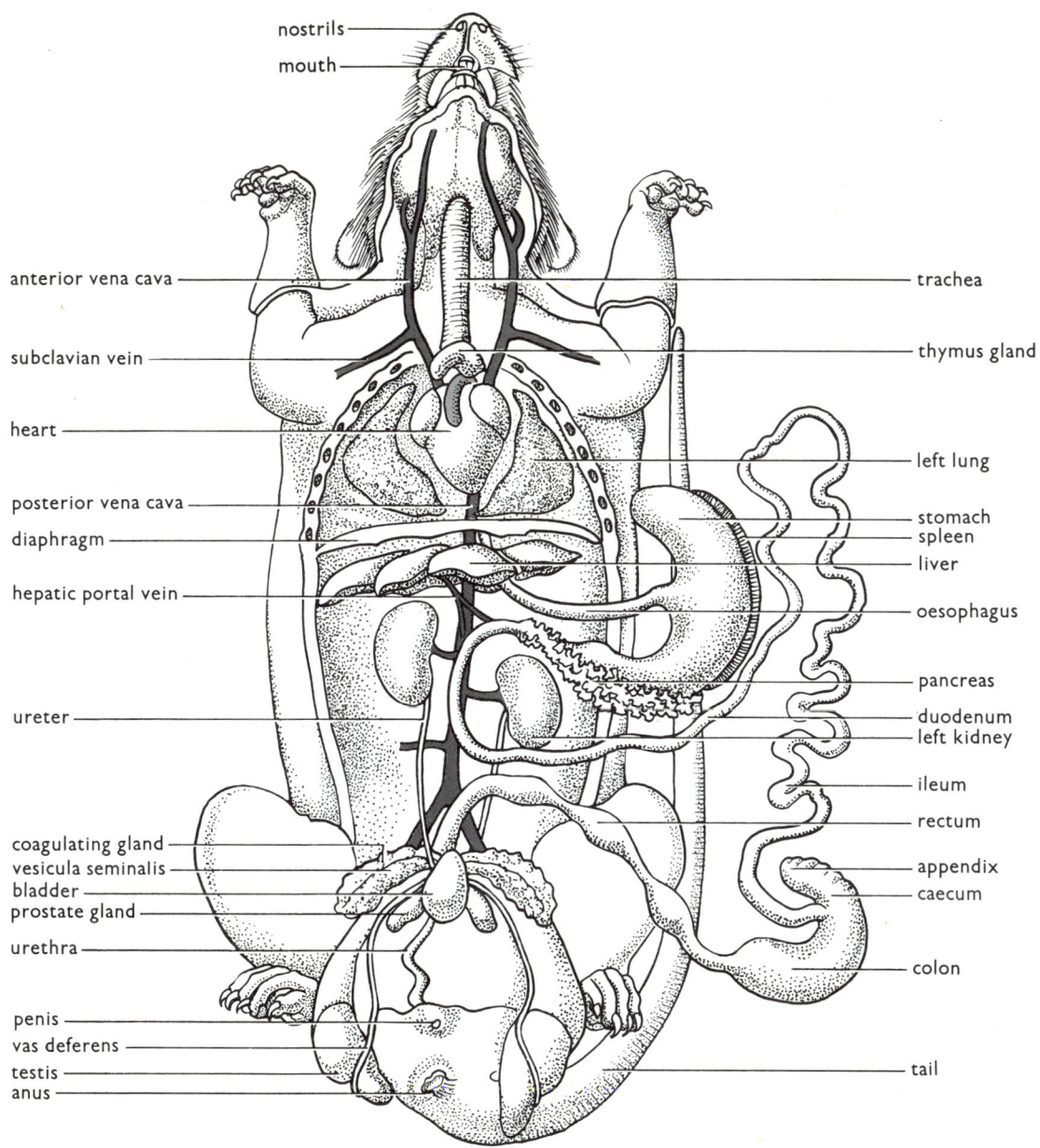

nostrils

mouth

anterior vena cava

subclavian vein

heart

posterior vena cava

diaphragm

hepatic portal vein

ureter

coagulating gland

vesicula seminalis

bladder

prostate gland

urethra

penis

vas deferens

testis

anus

trachea

thymus gland

left lung

stomach

spleen

liver

oesophagus

pancreas

duodenum

left kidney

ileum

rectum

appendix

caecum

colon

tail

RAT. DIAGRAM TO SHOW THE MAIN ORGANS AS REVEALED IN A DISSECTION FROM THE VENTRAL SURFACE

Lay the animal dorsal surface downwards on a dissecting board and stretch out the limbs. Identify its sex. Place through each foot a strong dissecting pin and slant it away from you so it will not get in the way during dissection. Lift the skin in the middle of the abdomen with a pair of forceps and snip through it with scissors. Make a cut lengthwise as far as the lower jaw in front and the reproductive opening behind. Loosen the skin from the body wall with your fingers and gently extend it, pinning it out to each side of the body. If necessary readjust the pins holding the feet to keep the limbs fully extended. Next cut into the body wall of the abdomen, cut forwards as far as the sternum in the thorax and backwards as far as the bladder. Make two lateral cuts at each end of your median cut and pin the body wall to the sides. Identify the main organs in the abdomen.

To dissect the thorax, make two lateral cuts through the ribs on each side, and carefully lift off the ventral portion of the thorax wall, freeing it from the diaphragm as you do this. Take care not to damage the heart. Identify the main organs of the thorax. At all stages of your dissection be careful not to dig the points of your scissors into the tissues, always keep the points cutting *towards* you.

The diagram above shows the principal structures inside the body of a male rat, as revealed in dissection. The limbs are drawn in their natural position (i.e. not pinned out as they would be during the course of dissection), and the digestive system has been displaced to the left side of the rat.

45

Nutrition

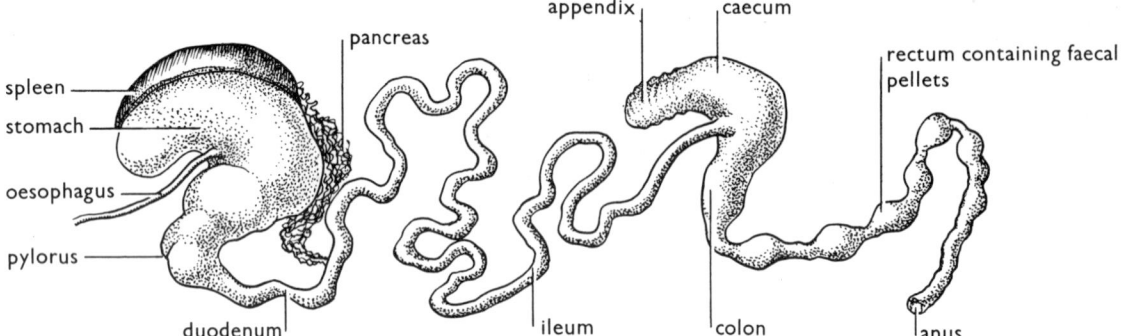

RAT, DIGESTIVE SYSTEM REMOVED FROM THE BODY

The rat is omnivorous.

The digestive system is suspended from the dorsal surface of the body cavity by means of a thin sheet of tissue called a *mesentery*. In this vessels of the blood and lymph systems travel to and from the digestive organs.

Having completed your general dissection, snip through the mesentery close to the rectum, and continue the process forwards as far as the stomach. Then cut through the rectum and oesophagus and lift the digestive system from the body, on to the dissecting board. Identify the main organs.

Teeth

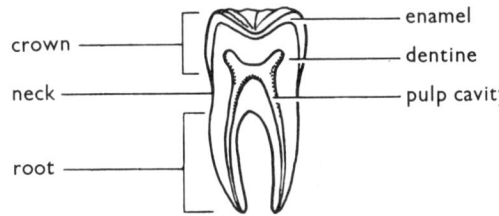

SECTION OF MAMMAL TOOTH

Mammalian teeth occur in sockets in the jaw bones. A typical tooth consists mainly of a hard substance *dentine*, which surrounds a central *pulp cavity* in which nerves and blood vessels are present. The outer surface or *crown* of the tooth is covered with a very hard and durable substance called *enamel*.

A tooth has three main regions: (1) the crown outside the gum; (2) the neck at the level of the gum; (3) the root which is fixed in the jaw socket by a cement substance.

Examine a tooth which has been split open along its length and identify the various structures and regions.

Mammalian teeth vary in shape and structure and are related to the type of food eaten. The main types are:

(a) *Incisors*, chisel shaped and used for cutting.
(b) *Canines*, pointed and used for tearing.
(c) *Premolars* and *Molars* used for grinding.

Incisors and canines have single roots, premolars and molars often have two or more roots.

Examine the skull of either a rabbit or a rat and identify the various teeth. In both these animals the incisors have roots which remain open at the base, and these teeth continue to grow throughout the life of the animal. Compare the inner

RABBIT SKULL

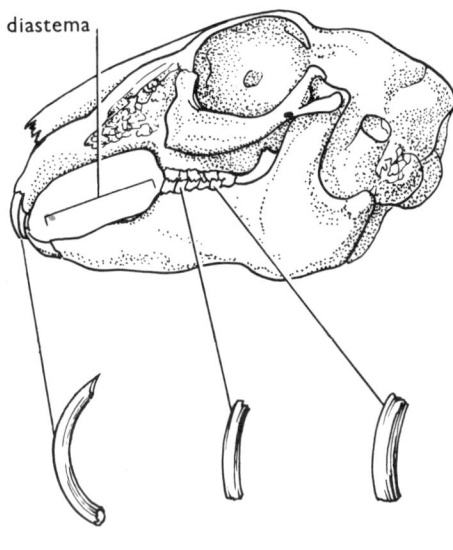

INCISOR   PREMOLAR   MOLAR

(concave) and the outer (convex) surface of these teeth. Often growth rings are visible on them and represent the daily amount of growth.

There are no canine teeth, and there is a wide gap in the jaw called the *diastema*, which lies between the incisor and molar teeth.

Dental formulae

$$\text{Rabbit} \quad i\frac{2}{1}: \quad c\frac{0}{0}: \quad pm\frac{3}{2}: \quad m\frac{3}{3}: \quad = 28.$$

$$\text{Rat} \quad i\frac{1}{1}: \quad c\frac{0}{0}: \quad pm\frac{0}{0}: \quad m\frac{3}{3}: \quad = 16.$$

Respiration

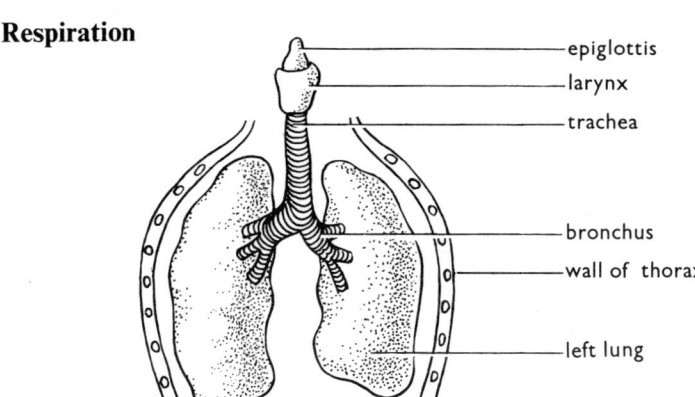

epiglottis
larynx
trachea
bronchus
wall of thorax
left lung
diaphragm

DIAGRAM OF RESPIRATORY SYSTEM

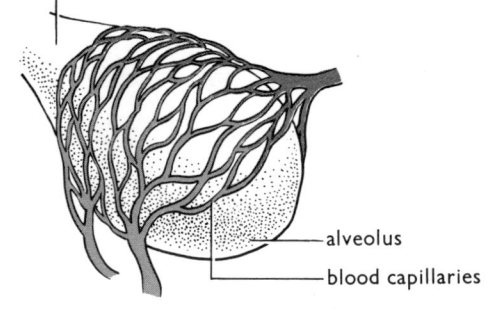

bronchiole
alveolus
blood capillaries

DIAGRAM OF ALVEOLUS,
HIGHLY MAGNIFIED

The respiratory system through which air passes to and from the lungs consists of: *nostrils* and *nasal cavity*. The *epiglottis*, a flap of skin at the back of the throat, which prevents food from entering the lungs during swallowing. The *larynx*, or voice-box. The *trachea* (wind-pipe), which is strengthened by rings of cartilage. This branches into two *bronchi* (tubes which open into finer tubes in the lung tissue).

The lungs are spongy structures, containing many small air

sacs called *alveoli*. In these, gases diffuse into and out of capillaries which lie very close to their walls. The lungs lie in the thorax on each side of the heart and are protected by membranes called *pleurae*.

Examine and identify the respiratory organs in a general dissection. (For experiments on the respiratory mechanism, see 21.1.)

Vascular System

VEINS

jugular
subclavian
anterior vena cava
right auricle
right ventricle
posterior vena cava
liver
hepatic
hepatic portal
intestine
renal

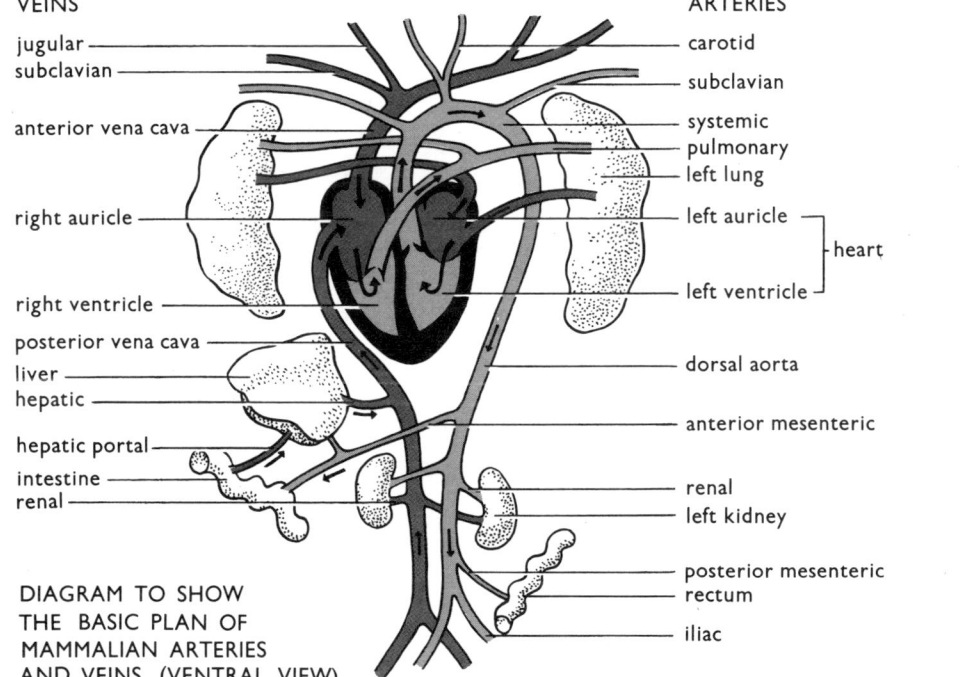

ARTERIES

carotid
subclavian
systemic
pulmonary
left lung
left auricle
left ventricle
heart
dorsal aorta
anterior mesenteric
renal
left kidney
posterior mesenteric
rectum
iliac

DIAGRAM TO SHOW
THE BASIC PLAN OF
MAMMALIAN ARTERIES
AND VEINS (VENTRAL VIEW)

DIAGRAM TO SHOW
SECTION OF ARTERY

DIAGRAM TO SHOW
SECTION OF VEIN

DIAGRAM TO SHOW
CAPILLARY NETWORK

This is a closed circuit of vessels, in which blood is circulated by the heart, and is the main transport system of the body.

Arteries are *thick-walled muscular* vessels, in which blood moves away from the heart.

Veins are *thin-walled* vessels, in which blood travels to the heart.

Capillaries are the connecting links between the arteries and veins occurring in the body tissues. They consist of fine tubes, often only wide enough to allow the passage of a thin stream of blood corpuscles. (See page 51.)

47

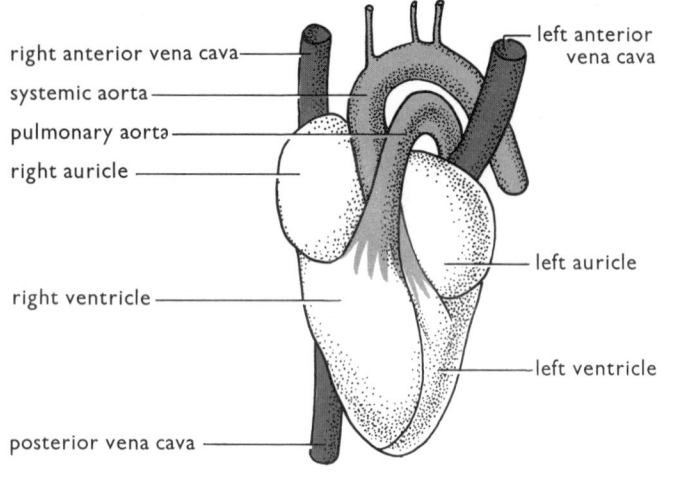

right anterior vena cava

systemic aorta

pulmonary aorta

right auricle

right ventricle

posterior vena cava

left anterior vena cava

left auricle

left ventricle

RAT HEART (VENTRAL VIEW)

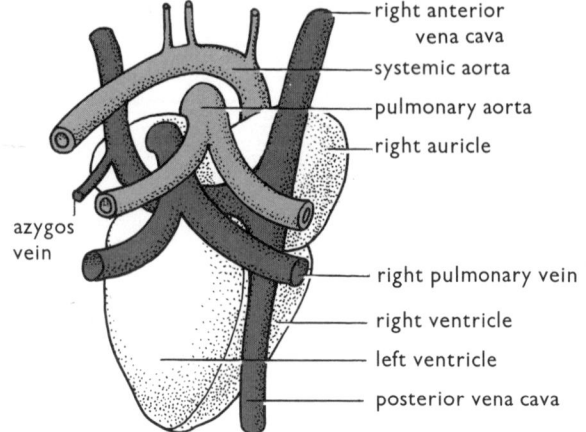

azygos vein

right anterior vena cava

systemic aorta

pulmonary aorta

right auricle

right pulmonary vein

right ventricle

left ventricle

posterior vena cava

RAT HEART (DORSAL VIEW)

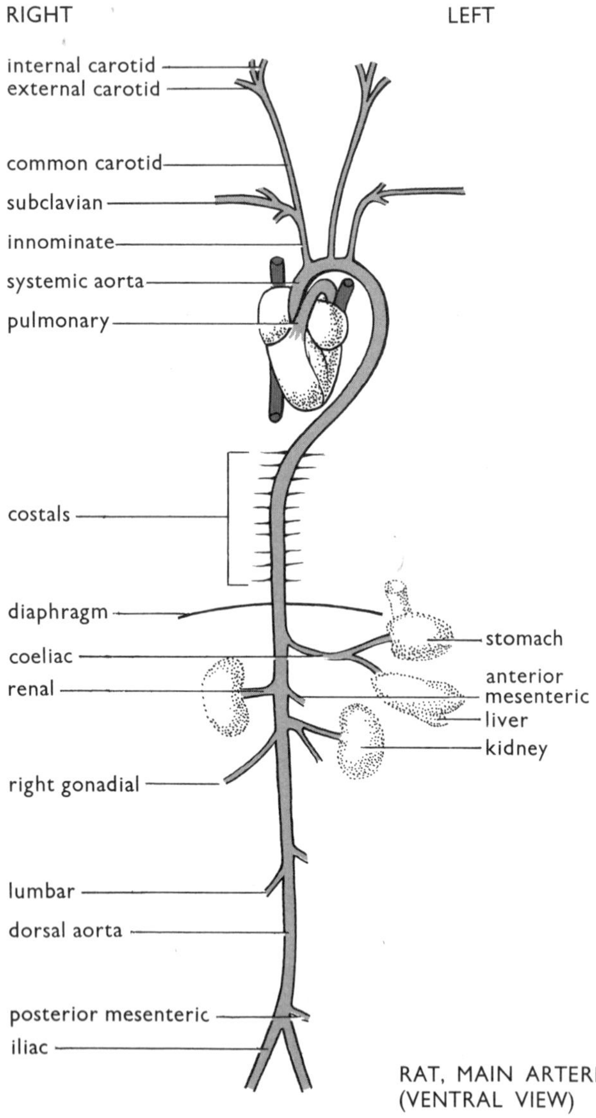

RIGHT   LEFT

internal carotid
external carotid

common carotid

subclavian

innominate

systemic aorta

pulmonary

costals

diaphragm

coeliac

renal

right gonadial

lumbar

dorsal aorta

posterior mesenteric

iliac

stomach

anterior mesenteric

liver

kidney

RAT, MAIN ARTERIES (VENTRAL VIEW)

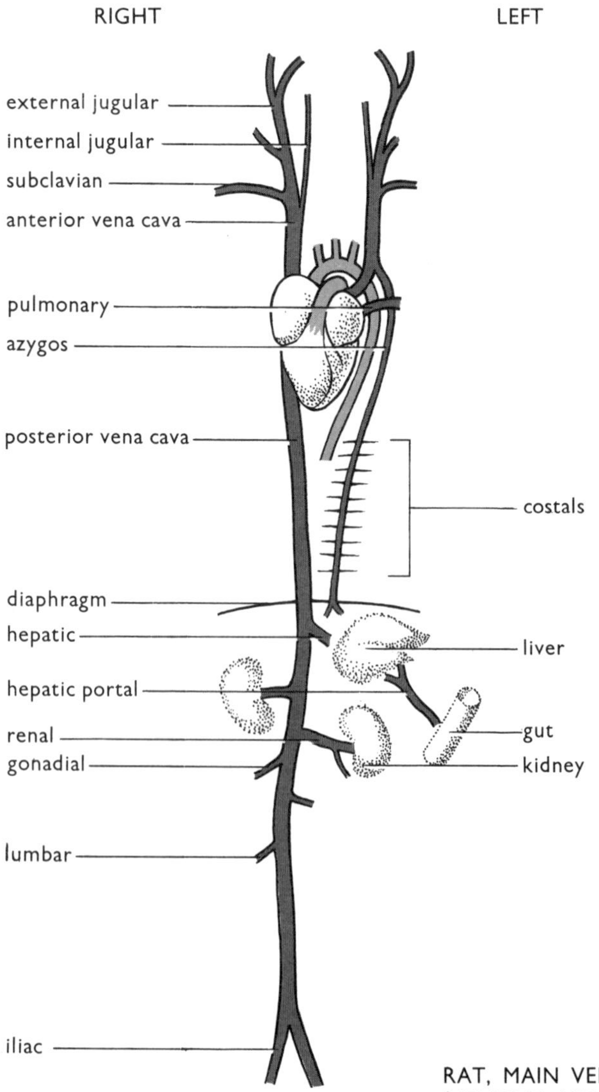

RIGHT   LEFT

external jugular

internal jugular

subclavian

anterior vena cava

pulmonary

azygos

posterior vena cava

costals

diaphragm

hepatic

hepatic portal

renal

gonadial

lumbar

iliac

liver

gut

kidney

RAT, MAIN VEINS (VENTRAL VIEW)

The Heart

The rat's heart lies in the ventral part of the thorax, slightly to the left of the mid-line, and between the two lungs. It is surrounded by a thin membrane called the *pericardium*. At the anterior end there are two *auricles*, each one a thin-walled compartment. At the posterior end there are two thick, muscular *ventricles*. Blood returning from the lungs enters the *left* auricle, and passes through into the *left* ventricle. Between the auricle and the ventricle there is a valve made from two flaps of skin. This called the *mitral valve*, and it prevents the blood from passing back into the auricle when the ventricle contracts. The *left* ventricle pumps blood into the main arteries of the body.

Blood returning from the body enters the *right* auricle, and passes through to the *right* ventricle. The opening between the two is protected by a valve made from three flaps of skin called the *tricuspid valve*.

During your general dissection examine the heart of a rat with a hand lens. Try to trace the path of the arteries and veins connected to it. In order to see the structure more clearly, dissect off the pericardium surrounding the heart.

Dissect a sheep's heart by making a slit through the ventral surface of the left and right ventricle, continuing your cut into the respective auricles. Study the internal structure, and note the arrangements of the valves, whose flaps are held in position by tendinous chords stretching from them into the ventricles.

The Arterial System

Identify the main arteries in your general dissection of the rat. The diagram on the opposite page (page 48) shows the position of the most important ones. Trace the path of the arteries systematically:

Carotid Arteries – serving the head and neck regions.
Subclavian Arteries – serving the fore limbs.
Systemic Aorta – the main artery in the thorax and abdomen.
Costal Arteries – serving the region of the ribs.
Coeliac and Mesenteric Arteries – serving the digestive system.
Renal Arteries – serving the kidneys.
Gonadial Arteries – serving the reproductive organs. (Note the different arrangement in male and female rats.)
Iliac Arteries – serving the region of the tail and hind limbs.
Pulmonary Artery – taking blood from the right ventricle to the lungs.

The Venous System

Identify the main veins in your general dissection.
Jugular Veins – in the region of the neck.
Subclavian Veins – in the region of the fore-limbs.
Anterior Venae Cavae and *Posterior Vena Cava* – which enter the heart by the right auricle.
Pulmonary Vein – entering the heart by the left auricle.
Azygos and Costal Veins – in the region of the thorax.
Hepatic Portal Veins – taking blood from the digestive system to the liver.
Hepatic Vein – taking blood from the liver to the Posterior Vena Cava.
Renal Veins – in the region of the kidneys.
Gonadial Veins – in the region of the reproductive organs.
Lumbar Veins – in the lumbar region of the abdomen.
Iliac Veins – in the region of the hind limbs and tail.

Excretion

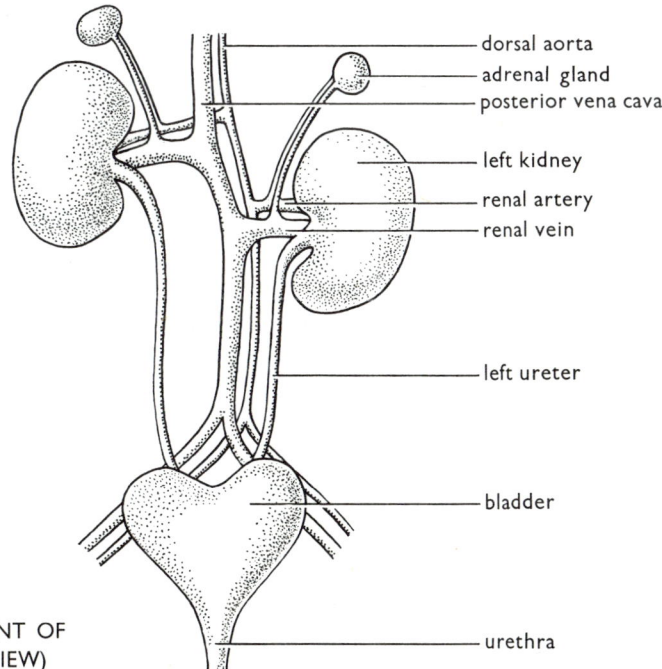

dorsal aorta
adrenal gland
posterior vena cava
left kidney
renal artery
renal vein
left ureter
bladder
urethra

DIAGRAM TO SHOW ARRANGEMENT OF EXCRETORY ORGANS (VENTRAL VIEW)

There are two kidneys in the dorsal region of the abdomen, the right slightly anterior to the left. Excretory substances filtered out of the renal blood supply, pass along *ureters* as *urine*, to be stored in the bladder. The urine then passes through a tube called the *urethra*, opening at the tip of the penis in the male, and near the vulva in the female. (See page 50.) Identify these organs in your general dissection.

49

Reproduction

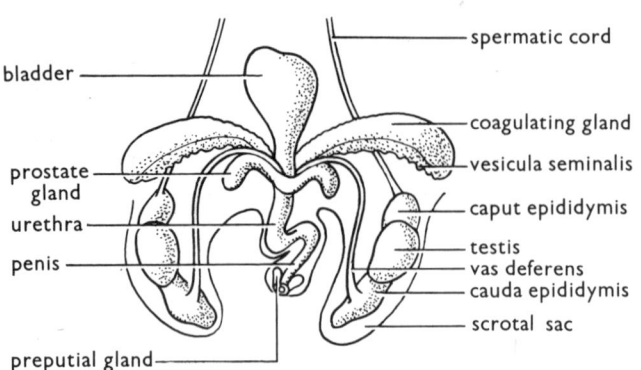

RAT, MALE REPRODUCTIVE SYSTEM
(VENTRAL VIEW)

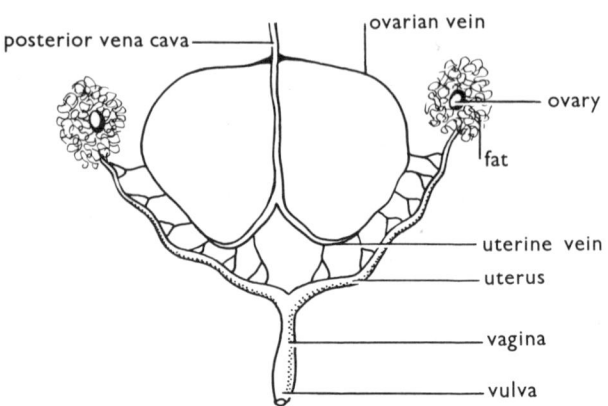

RAT, FEMALE REPRODUCTIVE SYSTEM
(VENTRAL VIEW)

Rat, male reproductive system   There are two testes inside a sheath of protective skin called the *scrotal sac*. Each testis has a *caput epididymis* and *cauda epididymis* on each side of it. Sperms produced in the testes are stored in the epididymes. The sperms reach the penis from each testis by passing along a duct called the *vas deferens*. Near the bladder there is a pair of prominent *vesiculae seminales* and *coagulating glands*, whose function is to produce a fluid which helps to keep the sperms alive. The spermatic cord contains the arteries and veins supplying the reproductive organs.

Identify the main organs in a general dissection of a male rat.

Rat, female reproductive system   The ovaries are small spherical bodies, about 3 mm in diameter, usually surrounded by fat, and in the region of the kidneys. Each ovary is connected to the *vagina* by means of a duct called the *uterus*. In a young female, each uterus is about 2 mm in diameter, but if the female is carrying embryos, these can be seen as swellings inside a greatly extended uterus. The arteries and veins supplying each uterus are prominent, and complete a circuit from the middle to the posterior part of the abdomen.

Identify the main organs in a general dissection of a female rat.

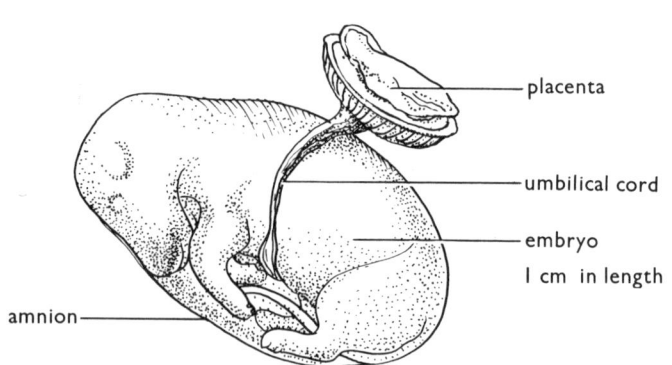

EMBRYO, WITH MEMBRANES AND PLACENTA

The embryo   This is the stage in development from the time an egg is fertilized until the young rat is born. The process takes about 22 days and is called the *gestation period*. During the dissection of the female rat embryos may be found in the uteri. Often as many as ten may be found. The diagram above shows a typical embryo after it has been removed from the uterus. It is surrounded by a protective membrane called the *amnion*, which encloses it in a fluid-filled cavity called the *amniotic cavity*. From the ventral region of the embryo's abdomen, the *umbilical cord* emerges. This contains arteries and veins belonging to the embryonic blood system and terminates in a button-shaped organ called the *placenta*. The placenta is attached to the wall of the mother's uterus and serves to supply the embryo with food and oxygen obtained from the mother's blood system. The embryo returns excretory substances and carbon dioxide via the placenta to the mother's blood. There is no direct contact between the blood systems of the embryo and the mother.

Identify the main structures in a young embryo by examining it with a hand lens.

50

13.6 MAMMAL-MICROSCOPIC STRUCTURE

13.6.1 Cells

These are microscopic units from which living things are made.

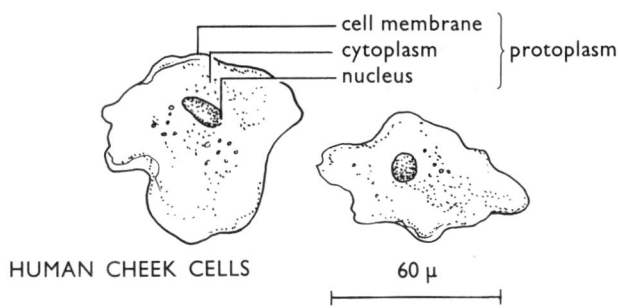

HUMAN CHEEK CELLS   60 μ

Human cheek cells

These are typical animal cells. They are made from protoplasm, a clear jelly-like material which is alive, and carries out all the living functions within its substance. The cell is bounded by a cell membrane, which separates it from its environment and encloses the other main parts—the cytoplasm and nucleus. The cytoplasm contains a number of granules, and towards the centre of it there is the nucleus, which is mainly responsible for the rate of reproduction of the cell.

Method of examination Take a clean spatula and scrape a little saliva from the inside of the cheek, placing the fluid on a slide. Cover with a coverslip, taking care to avoid the formation of air bubbles. Examine under high power, cutting the light down as much as possible. Draw a little weak iodine solution under the coverslip with a piece of blotting paper; it will make it easier to observe the cells, which are very transparent.

13.6.2 Tissues

These are groups of similar cells, joined together to perform a particular function.

EPITHELIAL TISSUES

These are layers of cells which cover the surface of the body or line spaces within it.

Cubical epithelium

Cells from the lining of a gland. Examine a prepared slide.

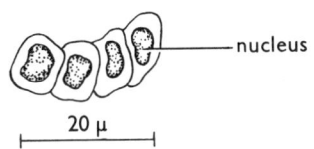

20 μ

Squamous epithelium

Inside the human cheek, the cells described above fit together rather like paving stones, to form a continuous tissue lining the inside of the cheek.

Columnar epithelium

Cells from the lining of the stomach (rabbit). Examine a prepared slide.

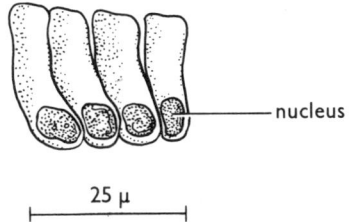

25 μ

CONNECTIVE TISSUES

These are groups of cells dispersed in a non-cellular matrix.

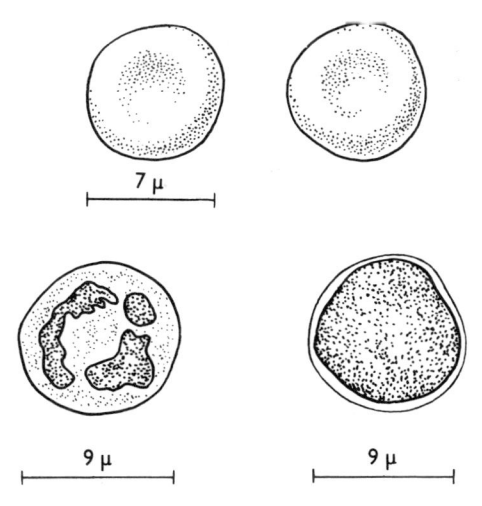

7 μ

9 μ   9 μ

POLYMORPHO - NUCLEOCYTE   MONOCYTE
(fragmented nucleus)   (with large nucleus)

Human blood

There is a fluid *matrix* or *plasma*, containing numerous cells.

Red corpuscles

Cells containing the respiratory pigment, *haemoglobin*. They have no nuclei, and in its place there is a depression in the centre of the cell. A healthy individual has some 5,000,000 per cu. mm.

White corpuscles

There are many types, identifiable by the appearance of their nuclei and the granular structure of the cytoplasm. A healthy individual has about 10,000 per cu. mm.

Method of examination Make a smear of human blood. (See 21.1 and 2.7.) Examine a prepared slide of mammalian blood, and compare it with frog's blood. (See page 36.)

Hyaline cartilage

This is the clear gristle at the ends of bones. (See page 40.) The *matrix* is transparent and contains many small cavities inside which are the cartilage cells. These are either single or, if they have recently reproduced, may be in groups of two or four. As they separate, they secrete new *matrix* around themselves.

Examine a prepared slide.

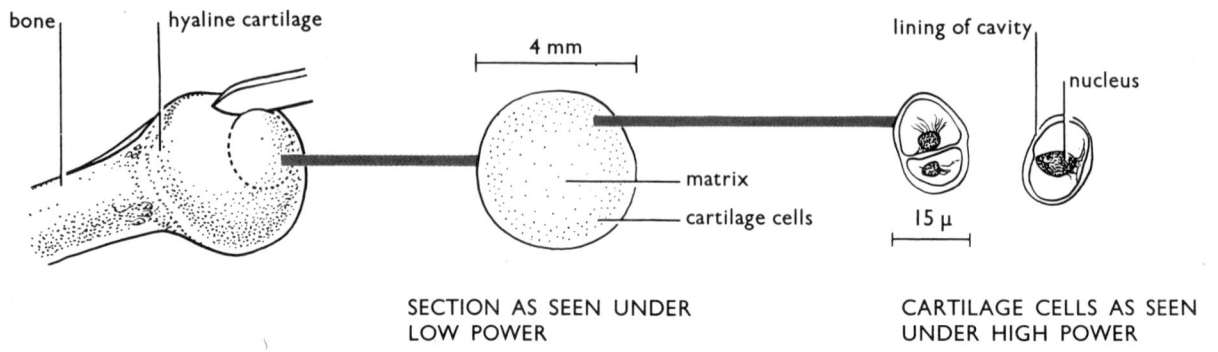

SECTION AS SEEN UNDER LOW POWER

CARTILAGE CELLS AS SEEN UNDER HIGH POWER

Bone

The microscopic structure of bone consists of concentric *haversian systems* of bone cells surrounding a central blood vessel. The bone cells are located in *lacunae* and they secrete the *matrix* of calcium salts. (See page 40.) Small *canaliculi* radiate on each side of each *lacuna*. The blood vessels are connected with the main blood system of the body, by capillaries passing through the *periosteum* surrounding the bone tissue.

Examine a prepared slide.

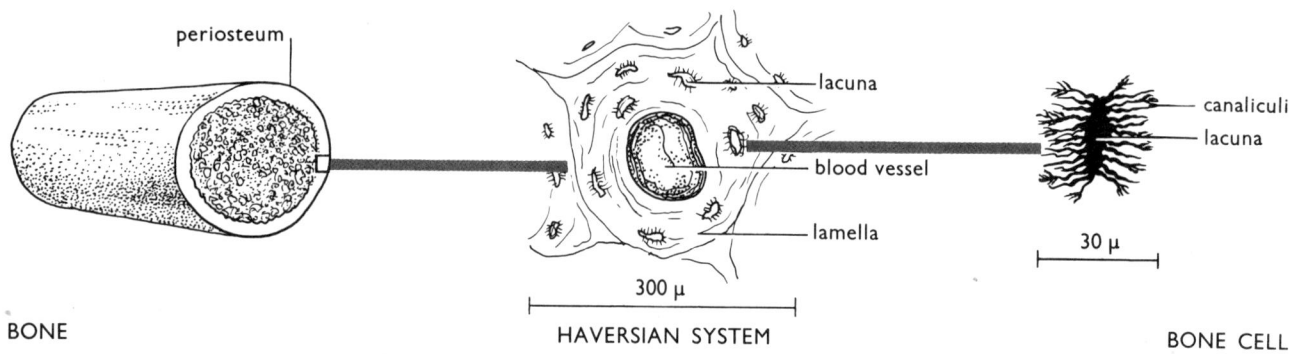

BONE

HAVERSIAN SYSTEM

BONE CELL

NERVOUS TISSUE

The cells of this tissue are responsible for conducting impulses from the sense organs to the muscular tissues of the body. A nerve cell (*neuron*) consists of a *cell body* containing the nucleus, and from it a long *process* emerges which carries the impulse.

Nerve cells are grouped together into nerves; tissues easily visible in dissections. A nerve consists of a sheath (*epineurium*) enclosing the nerve cells, which are gathered into distinct bundles, each surrounded by a *perineurium*.

Examine prepared slides of nervous tissue in which the *neurons* have been teased apart. Examine a transverse section of a nerve *process*, as seen for example in a section through a spinal cord.

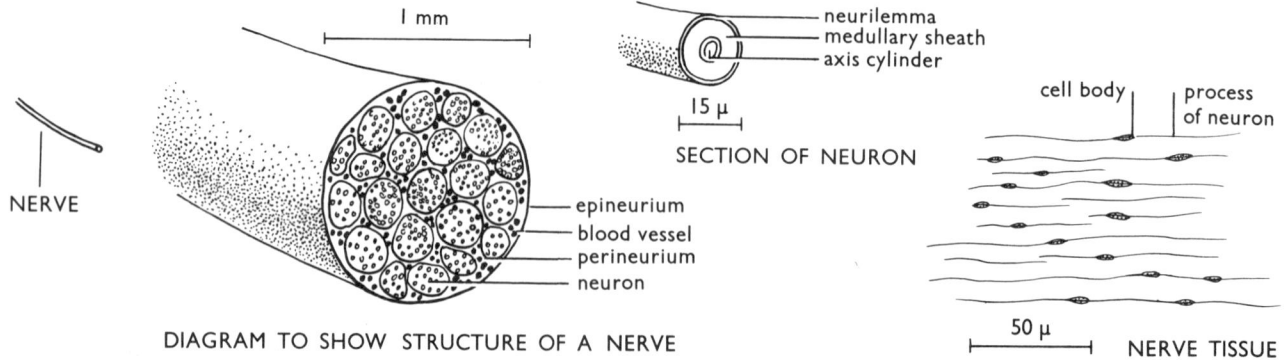

NERVE

DIAGRAM TO SHOW STRUCTURE OF A NERVE

SECTION OF NEURON

NERVE TISSUE

MUSCULAR TISSUES

These consist of elongated cells, able to change in length and so cause movement of the whole body.

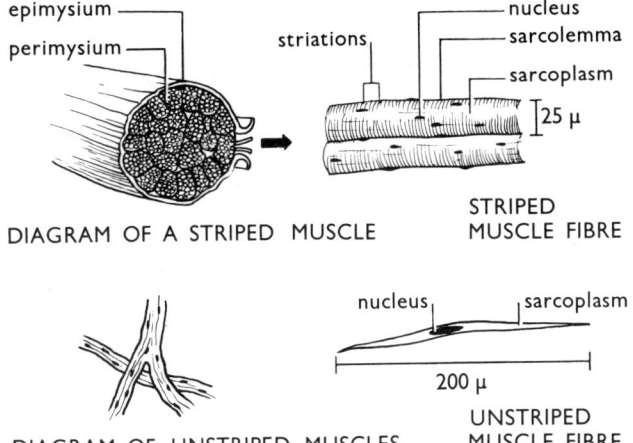

DIAGRAM OF A STRIPED MUSCLE

STRIPED MUSCLE FIBRE

DIAGRAM OF UNSTRIPED MUSCLES

UNSTRIPED MUSCLE FIBRE

Striped muscle

This consists of bundles of muscles cells surrounded by a connective tissue—the *perimysium*. Several groups of cells are in turn surrounded by an *epimysium*, to form a complete muscle. Muscles are attached to skeletal bones by *tendons* and are reponsible for the movement of the animal. Within the muscle tissue, capillaries and nerves occur. Each striped muscle cell or fibre is surrounded by a sheath—the *sarcolemma*, and includes a number of nuclei, a condition termed a *syncytium*. Numerous *striations* cross the *sarcoplasm* transversely, a characteristic feature of striped muscle fibres. Striped muscles are controlled voluntarily from the *Central Nervous System*.

Examine a prepared slide. Also tease up a small piece of fresh meat and examine under the high power.

Unstriped muscle

This consists of sheets of fibres, each one being a cell containing a single nucleus; there are no cross *striations*. Unstriped muscles are not controlled voluntarily and are served by the nerves of the *Autonomic Nervous System*.

Examine a prepared slide.

13.6.3 Organs

These are complete structures, consisting of a number of tissues associated for a particular function in the body.

The skin

This covers the outer surface of the body. It is waterproof; controls the body temperature by means of hair and fat as insulating devices preventing heat loss in cold surroundings. In warm conditions, excess heat is lost by the evaporation of sweat from sweat glands (not present in all mammals), which also excrete water containing excretory materials in solution. Nerves running in the deeper tissues of the skin are connected to surface sense organs detecting touch, temperature and pain.

Examine a prepared slide.

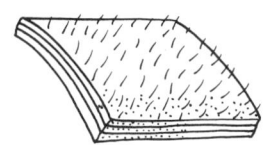

DIAGRAM OF SKIN

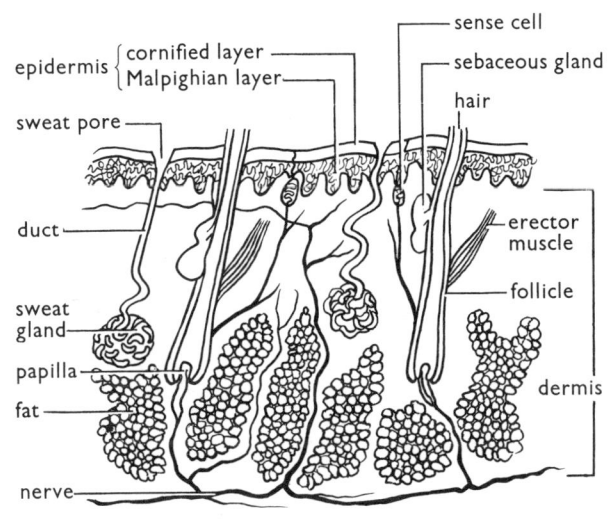

SECTION OF MAMMALIAN SKIN

The spinal cord

A transverse section shows the *meninges*, protective membranes covering the nervous tissue. The *grey matter* contains nerve cell bodies, whose *processes* are connected to *neurons* passing out in the ventral roots to the muscles. *Neurons* in the dorsal roots bring impulses in the spinal cord from the sense organs. *White matter* consists of *neurons* running longitudinally in the cord, to and from the brain.

Examine a prepared slide.

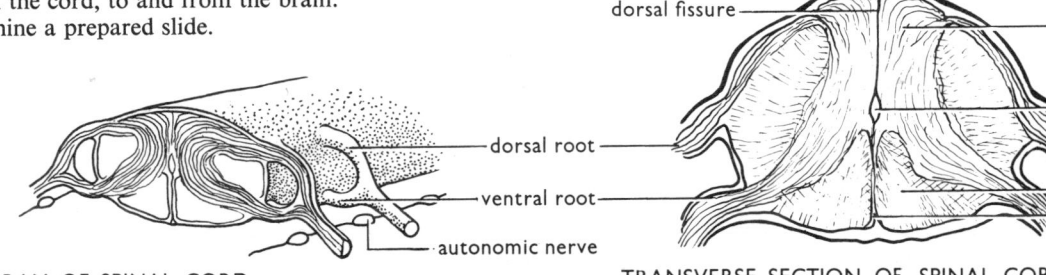

DIAGRAM OF SPINAL CORD

TRANSVERSE SECTION OF SPINAL CORD

53

The eye

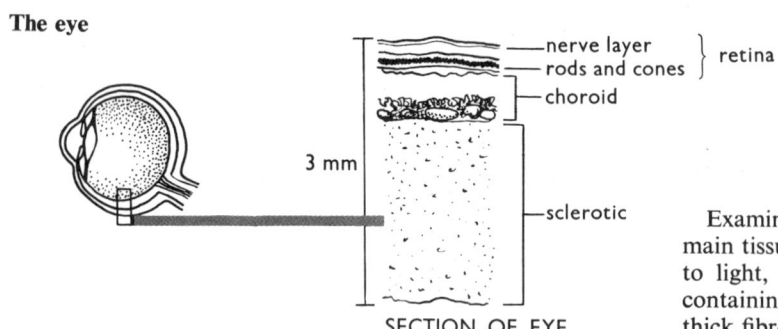

SECTION OF EYE

Examine a prepared slide of a section. There are three main tissues. The *retina*—containing rods and cones sensitive to light, and *neurons* from the optic nerve. The *choroid*—containing blood vessels and pigmented cells. The *sclerotic*—a thick fibrous tissue protecting the inner structures of the eye.

The ear

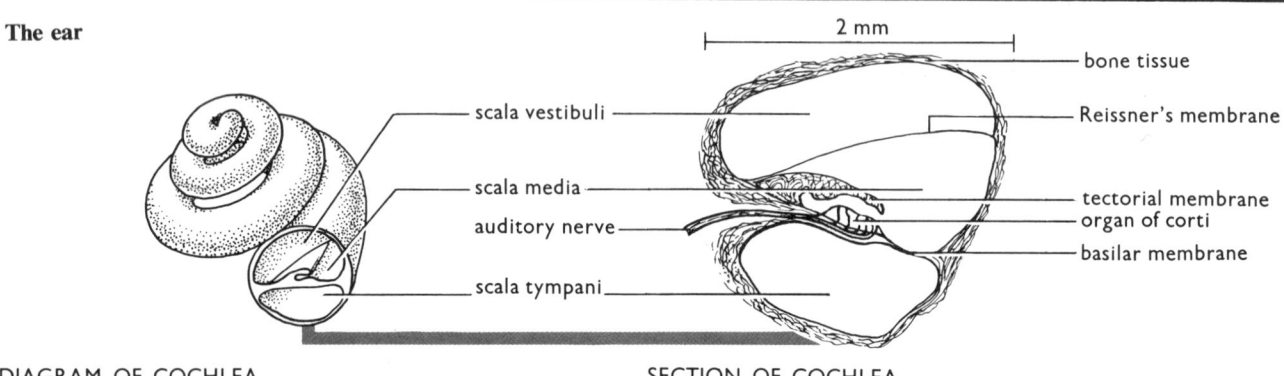

DIAGRAM OF COCHLEA   SECTION OF COCHLEA

Examine a prepared slide of a transverse section through the *cochlea*. In the centre of the *scala media* is a cavity containing *endolymph*, and the *organ of Corti* responsible for the detec-tion of sound waves. On each side of it is the *scala vestibuli* and the *scala tympani* containing *perilymph*.

The stomach

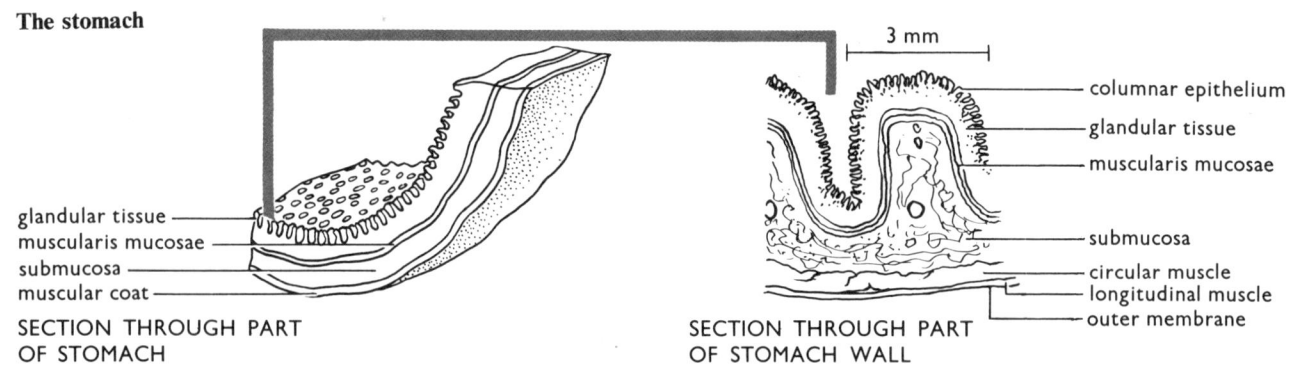

SECTION THROUGH PART OF STOMACH   SECTION THROUGH PART OF STOMACH WALL

Examine a prepared slide of a transverse section through a piece of stomach tissue. The cavity of the stomach is lined by columnar epithelium, together with glandular tissue responsible for the secretion of gastric juice. The surface of the stomach is covered with a thick layer of unstriped muscular tissue.

The intestine

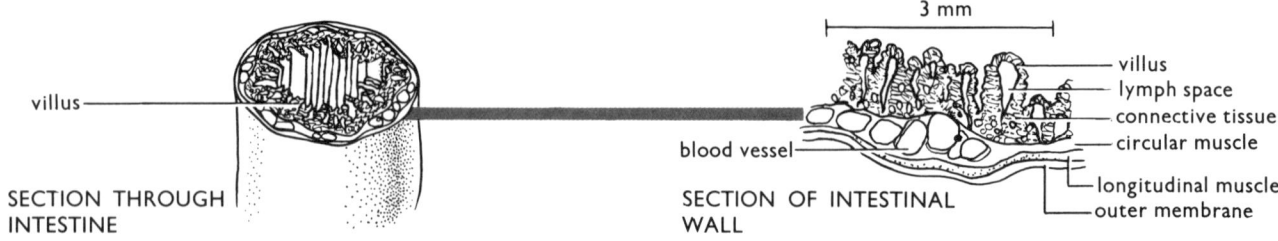

SECTION THROUGH INTESTINE   SECTION OF INTESTINAL WALL

Examine a prepared slide of a section through a piece of intestine. The cavity is lined by epithelial tissue, thrown into folds—the *villi*. Each *villus* contains blood vessels and a lymph space. The intestine is covered by a thin layer of unstriped muscular tissue.

The kidney

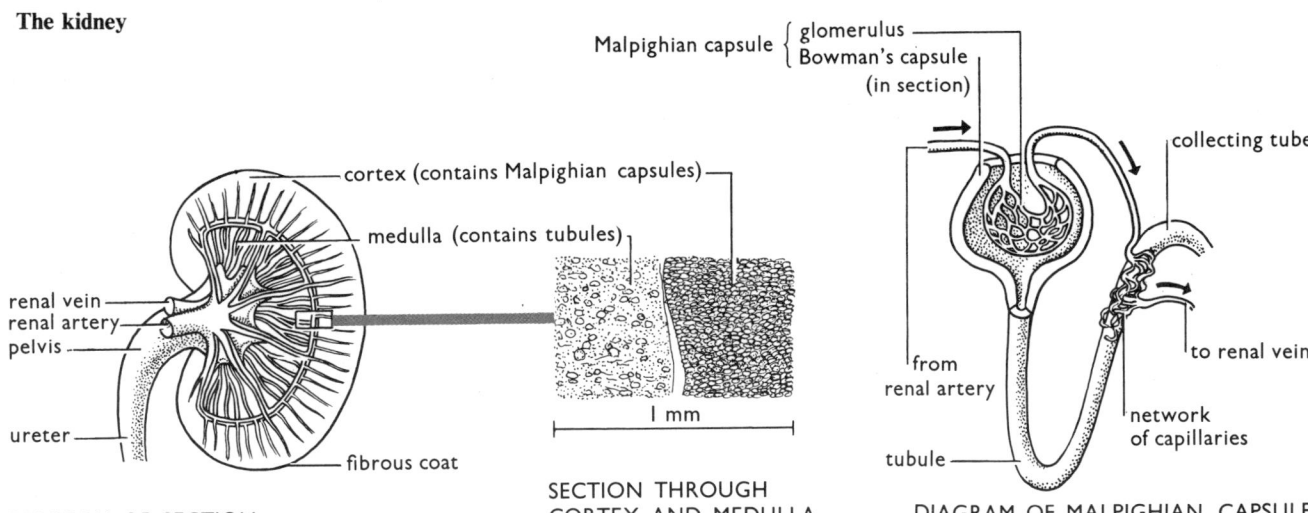

Malpighian capsule { glomerulus
Bowman's capsule
(in section)

collecting tube

cortex (contains Malpighian capsules)

medulla (contains tubules)

renal vein
renal artery
pelvis

ureter

fibrous coat

from renal artery

to renal vein

network of capillaries

tubule

1 mm

DIAGRAM OF SECTION
THROUGH KIDNEY

SECTION THROUGH
CORTEX AND MEDULLA

DIAGRAM OF MALPIGHIAN CAPSULE

This consists of an outer cortex and an inner medulla. Blood is conveyed from the renal artery into the tissue of the cortex, where excretory products are filtered out of it by the *malpighian capsules*. The urine travels from the capsule through tubules which connect together from other capsules in the cortex, and then through the medulla. The urine finally leaves the kidney via its pelvis, entering the ureter. The filtered and purified blood leaves the kidney tissue in the renal vein.

Examine a prepared slide.

The ovary

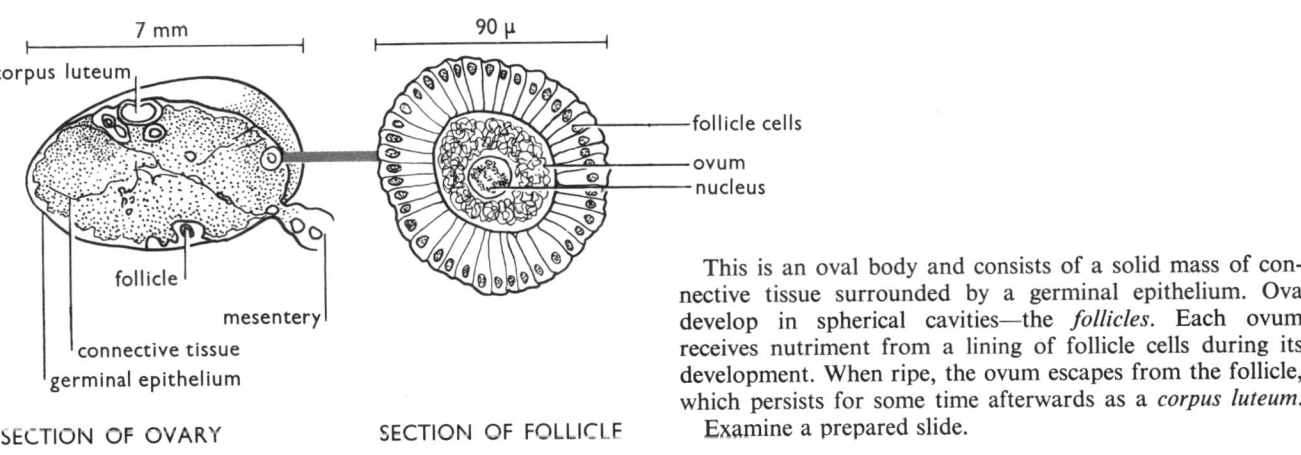

7 mm

90 μ

corpus luteum

follicle cells
ovum
nucleus

follicle

mesentery

connective tissue
germinal epithelium

SECTION OF OVARY

SECTION OF FOLLICLE

This is an oval body and consists of a solid mass of connective tissue surrounded by a germinal epithelium. Ova develop in spherical cavities—the *follicles*. Each ovum receives nutriment from a lining of follicle cells during its development. When ripe, the ovum escapes from the follicle, which persists for some time afterwards as a *corpus luteum*.

Examine a prepared slide.

The testis

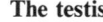
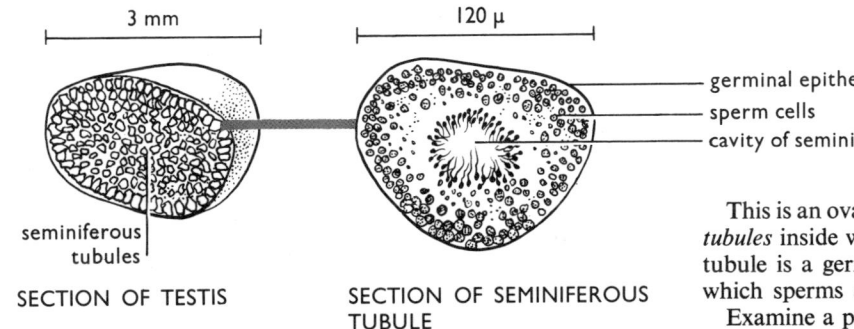

3 mm

120 μ

germinal epithelium
sperm cells
cavity of seminiferous tubule

seminiferous tubules

SECTION OF TESTIS

SECTION OF SEMINIFEROUS
TUBULE

This is an oval body, containing a mass of small *seminiferous tubules* inside which the sperms are formed. The lining of the tubule is a germinal epithelium producing sperm cells, from which sperms are formed towards the centre of the tubule.

Examine a prepared slide.

55

III BOTANY

The Plant Kingdom consists of two main groups.

Non-vascular plants These have no conducting tissues.

(i) BACTERIA and FUNGI have no chlorophyll, and obtain their food parasitically or saprophytically.

(ii) ALGAE have chlorophyll. Some are unicellular (e.g. *Pleurococcus* spp.); others are multicellular either in the form of cylindrical filaments (e.g. *Spirogyra* spp.) or a ribbon-shaped THALLUS (e.g. many Seaweeds).

(iii) BRYOPHYTA (Mosses) and HEPATOPHYTA (Liverworts). Terrestrial and contain chlorophyll. Their life cycle includes a GAMETOPHYTE generation inhabiting moist places alternating with a SPOROPHYTE generation.

Vascular plants These have conducting tissues carrying fluids within their leaves, stems and roots. The majority contain chlorophyll and live on land.

(i) SEEDLESS PLANTS. They have a life cycle similar to Mosses and Liverworts.

    ARTHROPHYTA (Horsetails)

    PTEROPHYTA (ferns)

(ii) SEED-BEARING PLANTS. These do not require water in the habitat for fertilization. Seeds are produced which can lie dormant until conditions favour germination.

    CONIFEROPHYTA. Coniferous plants.

    ANTHOPHYTA. Flowering plants.

In the classification of the Plant Kingdom, the term *Division* is most frequently used, and is synonymous with the term *Phylum* used in the classification of the Animal Kingdom.

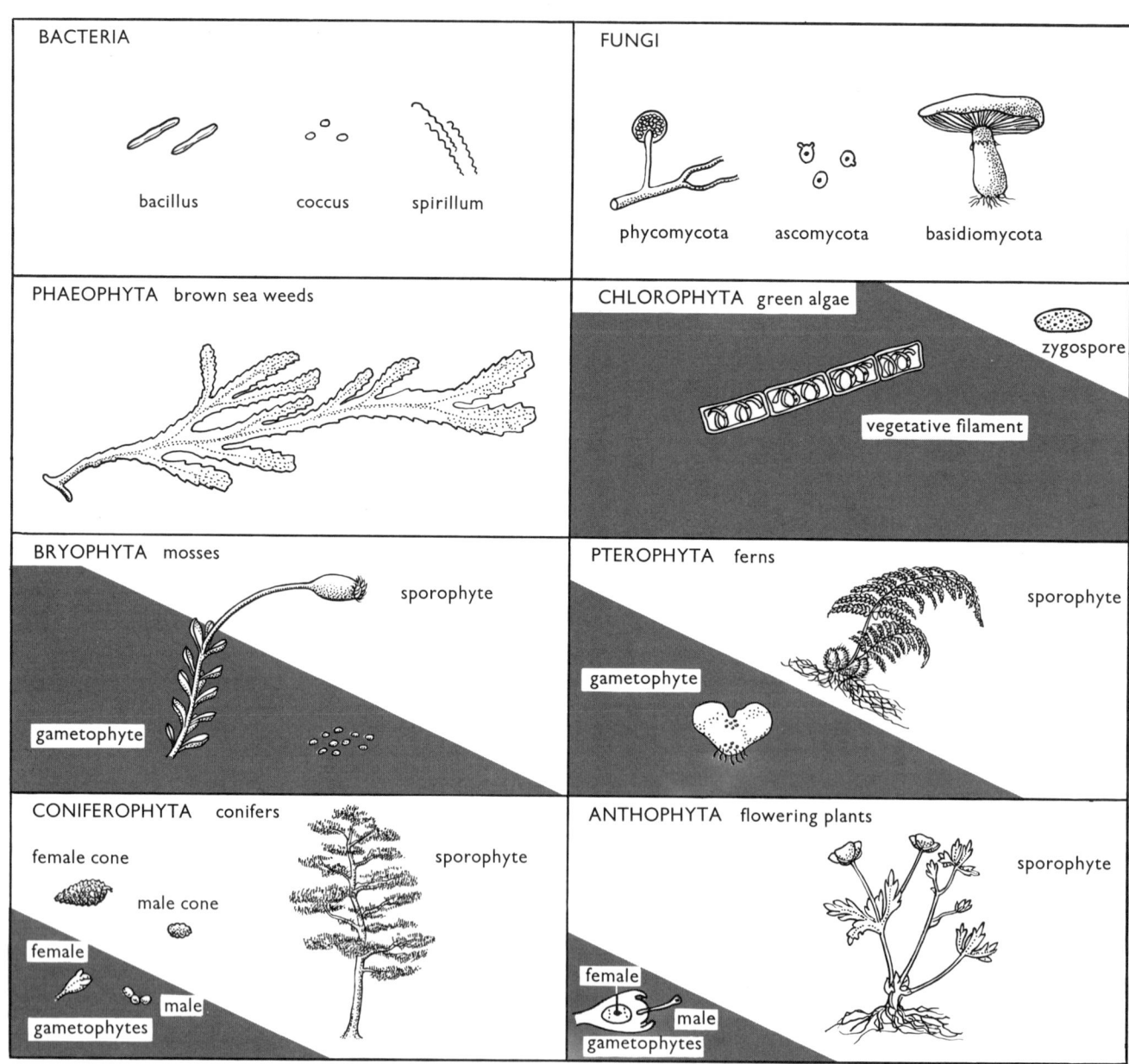

BACTERIA — bacillus, coccus, spirillum

FUNGI — phycomycota, ascomycota, basidiomycota

PHAEOPHYTA brown sea weeds

CHLOROPHYTA green algae — zygospore, vegetative filament

BRYOPHYTA mosses — sporophyte, gametophyte

PTEROPHYTA ferns — sporophyte, gametophyte

CONIFEROPHYTA conifers — female cone, male cone, sporophyte, female, male, gametophytes

ANTHOPHYTA flowering plants — sporophyte, female, male, gametophytes

14 Non-vascular Plants

Plants without water-conducting vessels.

14.1 CHLOROPHYTA

Green algae, simple plants containing chloroplasts. Unicellular; or multicellular in the form of threads.

14.1.1 *Pleurococcus* spp.

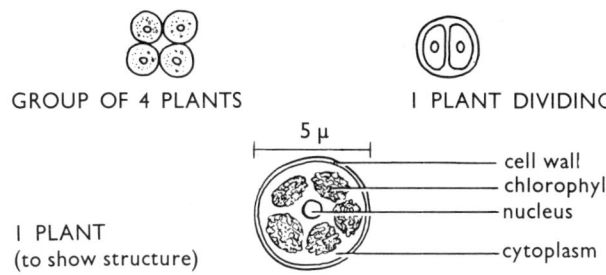

GROUP OF 4 PLANTS   I PLANT DIVIDING

I PLANT
(to show structure)

— cell wall
— chlorophyll
— nucleus
— cytoplasm

It occurs as a fine green powder on tree trunks and fences. It is unicellular, each cell surrounded by a spherical cell wall, and containing a nucleus and *chloroplast*. It reproduces by binary fission and recently divided cells can often be seen in groups.

Examine, by placing a little of the powder in a drop of water on a slide, and squash under a coverslip.

14.1.2 *Euglena* spp.

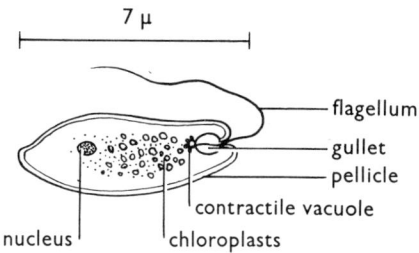

— flagellum
— gullet
— pellicle
— contractile vacuole
nucleus   chloroplasts

A unicellular organism, which moves rapidly by means of a *flagellum*, a structure which draws it through water by a screw-like action. There is a nucleus and *chloroplast*, and it is positively phototactic, moving towards bright light. It reproduces by binary fission, splitting along its length.

Examine by placing a drop of culture on a slide, and covering with a coverslip. Examine under low and high power. In a short time many specimens will slow up, making examination easier.

14.1.3 *Spirogyra* spp.

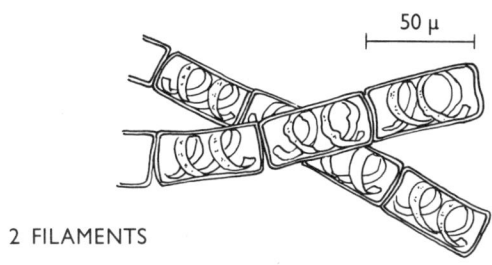

50 μ

2 FILAMENTS

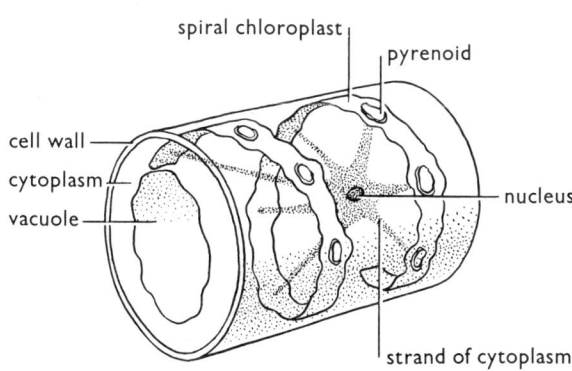

spiral chloroplast
pyrenoid
cell wall
cytoplasm
vacuole
nucleus
strand of cytoplasm

DIAGRAM OF I CELL (to show structure)

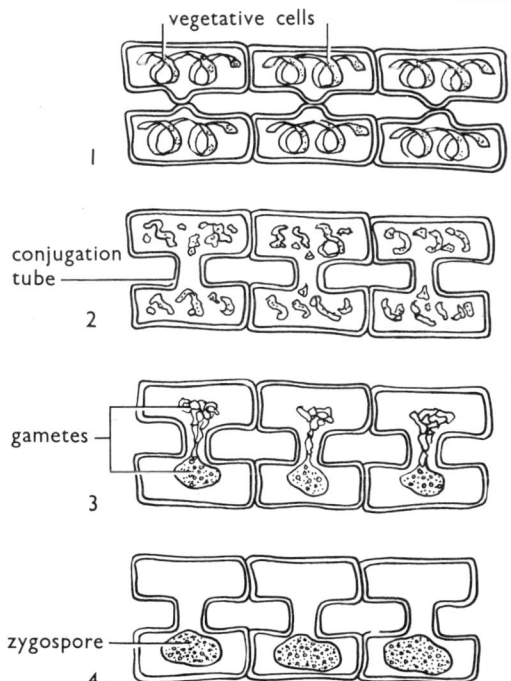

vegetative cells

I

conjugation tube

2

gametes

3

zygospore

4

4 STAGES IN SEXUAL REPRODUCTION

Habitat and structure A fresh-water alga consisting of thin cylindrical filaments, made from cells placed end to end. Each has a cellulose cell wall lined with *cytoplasm*. A central *vacuole* has a nucleus in it, supported by strands of *cytoplasm* reaching out to the margin. A spiral-shaped *chloroplast* in the rind of *cytoplasm* contains red *pyrenoids* round which starch is stored.

Reproduction It reproduces vegetatively; single cells break off, growing into new filaments. Sexual reproduction occurs when the cell substances of adjacent cells in two parallel filaments pass through *conjugation tubes* and fuse. This is *conjugation*, producing a resting spore—the *zygospore*.

Method of examination Examine fresh filaments in a drop of water on a slide. (See Photosynthesis experiments 20.2.2; also 23.3.2.) Examine prepared slides showing conjugation.

14.2 PHAEOPHYTA

Brown Algae, marine seaweeds.

14.2.1 Common Wrack (*Fucus serratus*)

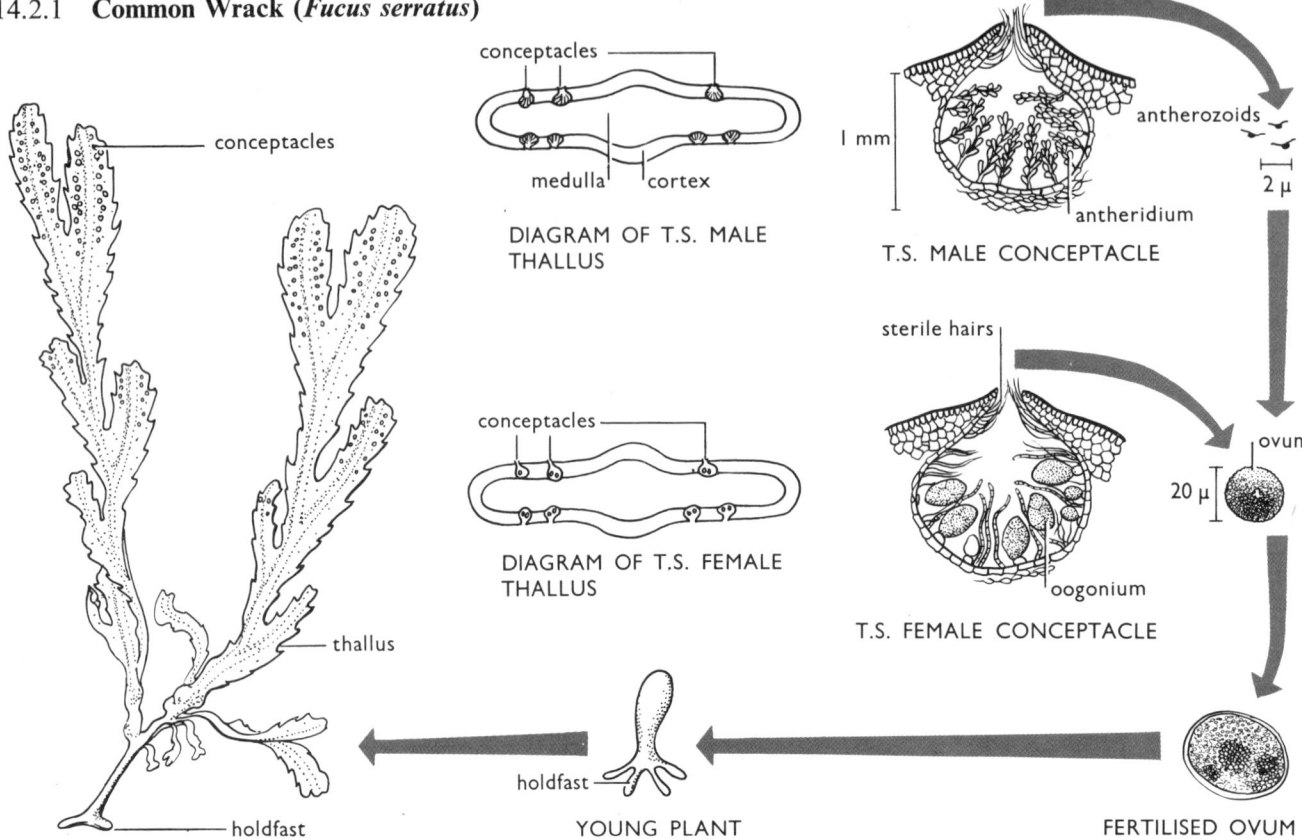

conceptacles

DIAGRAM OF T.S. MALE THALLUS

medulla | cortex

T.S. MALE CONCEPTACLE

antherozoids

2 μ

1 mm

antheridium

conceptacles

DIAGRAM OF T.S. FEMALE THALLUS

sterile hairs

T.S. FEMALE CONCEPTACLE

oogonium

ovum

20 μ

thallus

holdfast

YOUNG PLANT

holdfast

FERTILISED OVUM

WHOLE PLANT

A common seaweed found near the low tide mark. The flat thallus is attached to a rock by a holdfast organ. Growth takes place from an apical cell, forming two branches of equal size either side of it. The reproductive organs occur at the tip of the thallus; male *conceptacles* produce *antheridia* which form *antherozoids*. Female *conceptacles* are on a different plant, and have *oogonia* in which *ova* develop. Fertilization takes place in the sea, ripe gametes (*ova* and *antherozoids*) are shed into it on the incoming tide. A ripe egg becomes attached to a rock and grows into a new plant.

Examine fresh seaweed. Squeeze some of the slime from *conceptacles* on to a slide, and look for gametes (see 23.3.3). Examine prepared slides of transverse sections through the region of the reproductive organs. Extract jelly from seaweed by boiling a thallus in water and filter through a small mesh sieve. (See 22.2.2.)

14.2.2 Other Algae (Marine forms)

Phaeophyta. Brown Algae

Rhodophyta. Red Algae

Chlorophyta. Green Algae

Knotted Wrack
(*Ascophyllum nodosum*)

Carrageen
(*Chondrus crispus*)

Coralline
(*Corallina officinalis*)

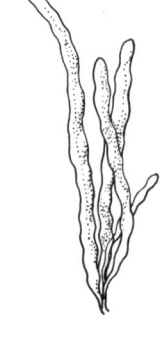

Enteromorpha spp.

14.3 SCHIZOMYCOTA

Bacteria; microscopic organisms which obtain their food saprophytically and parasitically.

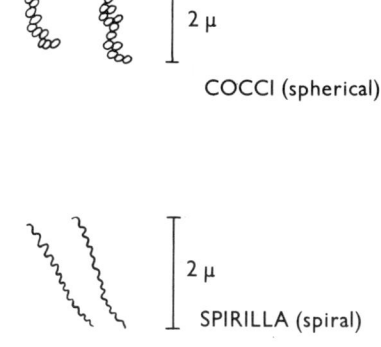

COCCI (spherical)

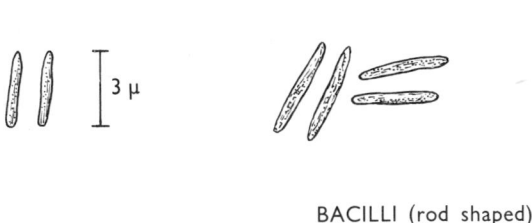

BACILLI (rod shaped)

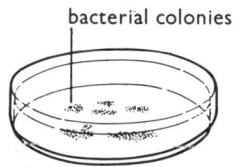

SPIRILLA (spiral)

Habitat and structure Bacteria are minute organisms less than 8 μ in length. They have a firm membrane surrounding protoplasm, but no obvious nucleus. They are parasites or saprophytes, reproducing rapidly by binary fission in suitable food material. When there is none available, bacteria form spores which survive for long periods.

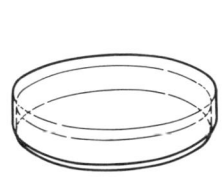

I. STERILISED PETRI DISH CONTAINING AGAR

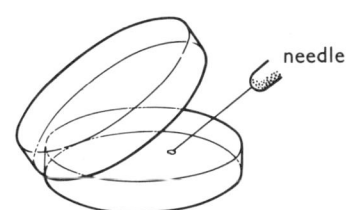

2. ADDING A SPECK OF SOIL WITH A STERILISED NEEDLE

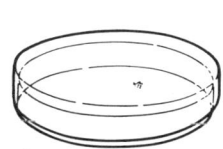

3. CLOSED PETRI DISH LEFT FOR 48 HOURS

4. BACTERIAL COLONIES DEVELOPING

Method of examination Sterilize a number of clean petri dishes by heating in an oven at 120°C for an hour. Prepare some agar jelly by adding 3 g agar to 100 cm³ water, to which a little boiled potato has been added. Boil over a water bath. If necessary, the boiling agar jelly can be filtered by passing it through a warm filter funnel containing a little cotton wool. Boil again after filtering, and then pour into petri dishes so that the bottom is just covered. Allow the dishes to cool to room temperature. Then expose different dishes for a few seconds by coughing, adding dust, soil, etc. Leave in a warm room for 48 hours. Bacterial cultures which develop on the agar will be visible as small clear blotches. Remove a small quantity with a needle and place on a slide in a drop of water. Examine under low and high power, using a minimum amount of light.

Examine bacteria from fresh sources, i.e. water in which flies, beans, hay, etc., have been allowed to decay. Many motile forms, especially spirilla, may be seen in this way.

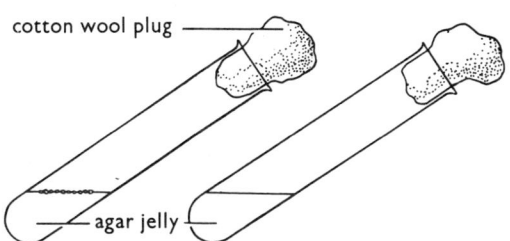

Experiments on control of bacteria

(a) *Disinfectants* Prepare potato agar as described above. Sterilize several boiling tubes and in each put 1 ml and ½ ml of various antiseptics and disinfectants. Pour hot freshly prepared agar jelly into the tubes, plug with cotton-wool which has been briefly passed through a flame to sterilize it, and allow each tube to cool in a slanting position. Transfer bacterial cultures into these tubes using a sterilized needle. After removing the cotton-wool, flame it once again before using it to plug the tube. Record the effectiveness of different disinfectants in varying strengths in preventing the development of bacteria.

(b) *Antibiotics* Obtain commercial preparations of antibiotics, and test their effectiveness in controlling bacterial cultures growing in petri dishes.

59

14.4 PHYCOMYCOTA

Fungi; plants without chlorophyll which obtain their food saprophytically or parasitically.

14.4.1 The Pin Mould (*Mucor hiemalis*)

Habitat It grows on moist organic matter, and will grow on a piece of moist bread kept under a bell jar for a few days.

Structure The fungus consists of minute branching tubular *hyphae* interwoven into a mesh—the *mycelium*. The tips of the *hyphae* penetrate the bread and convert it into soluble substances which can be absorbed by the fungus.

Asexual reproduction *Sporangiophores* grow vertically upwards from the *mycelium*, developing black dots at their tips. These are *sporangia*, inside which many multinucleate spores develop. The *sporangia* burst when ripe, setting free light spores which are dispersed by the wind.

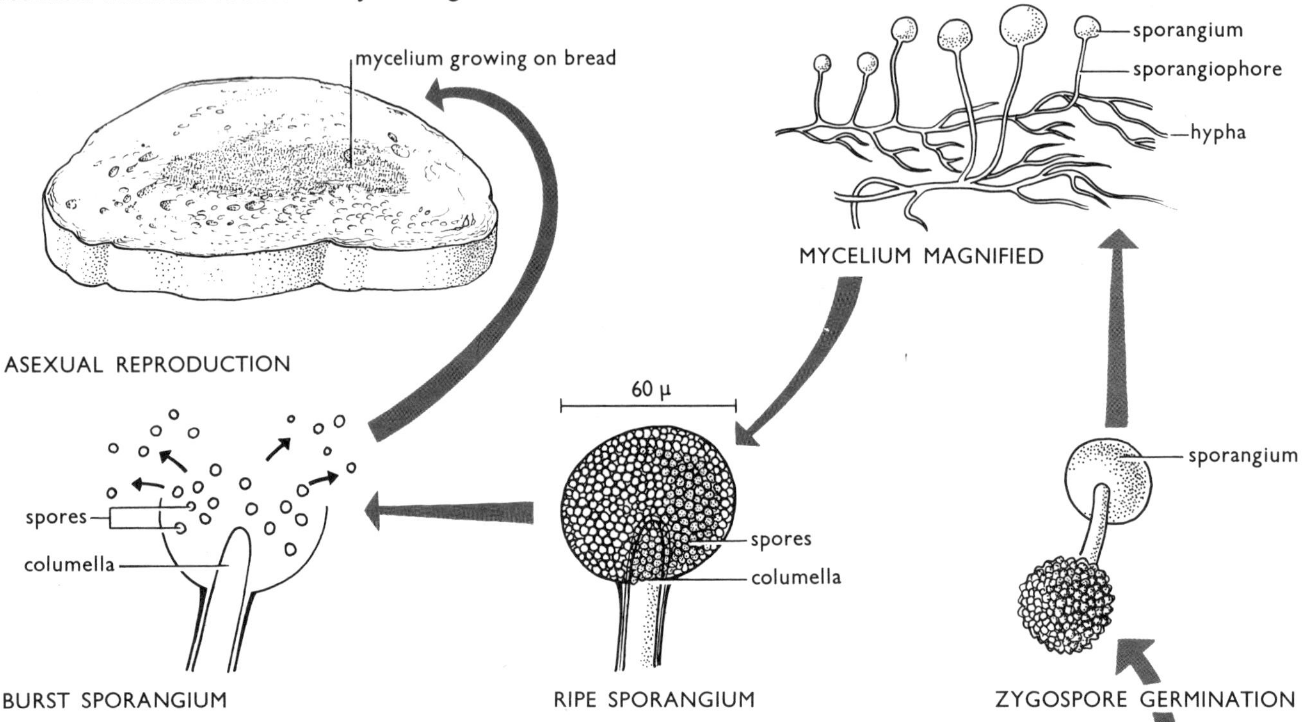

mycelium growing on bread

ASEXUAL REPRODUCTION

sporangium
sporangiophore
hypha

MYCELIUM MAGNIFIED

60 μ

spores
columella

BURST SPORANGIUM

spores
columella

RIPE SPORANGIUM

sporangium

ZYGOSPORE GERMINATION

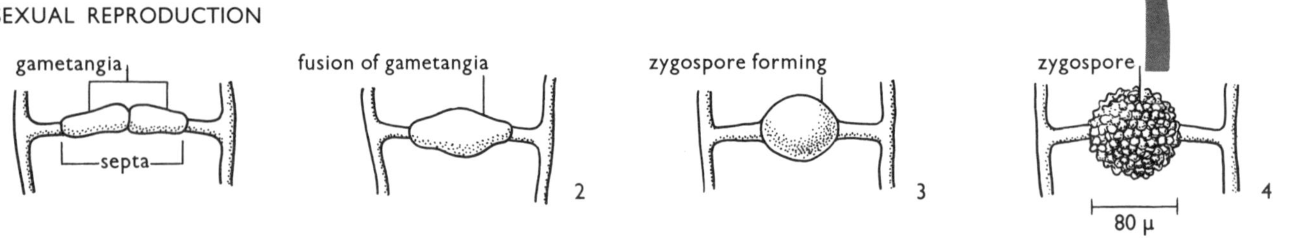

SEXUAL REPRODUCTION

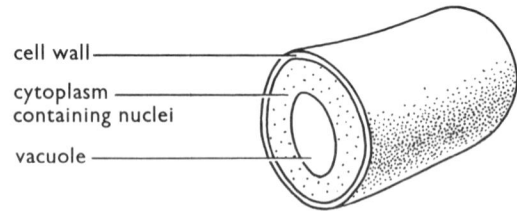

gametangia
septa

fusion of gametangia

2

zygospore forming

3

zygospore

80 μ

4

Sexual reproduction An outgrowth appears from each of two adjacent *hyphae*, they join at their tips, swell and separate from the rest of the *hyphae* by septa. Several nuclei are trapped in each swelling—the *gametangium*. Shortly afterwards the wall between each *gametangium* dissolves and the nuclei then fuse in pairs. A thick outer wall forms and a large warty *zygospore* develops. This can remain dormant for several months. Under favourable conditions it germinates to produce one *sporangium* containing uninucleate spores. These are liberated and each one germinates, if it falls on suitable food, to form a new *mycelium*.

Method of examination Pick a small piece of *mycelium* together with *sporangia* from a piece of bread and place it on a slide in a drop of water. Cover with a coverslip and examine under low and high power.

Examine a prepared slide showing *zygospores* and *gametangia*.

cell wall
cytoplasm containing nuclei
vacuole

DIAGRAM OF HYPHA

14.4.2 *Saprolegnia* **spp.**

This fungus is often responsible for disease in fresh-water fish. Its *hyphae* may be found on the gills; also on the bodies of dead flies. Examine a piece of *mycelium* in a drop of water on a slide. *Zoospores* can often be seen moving inside the *sporangia*. (See 23.3.1.)

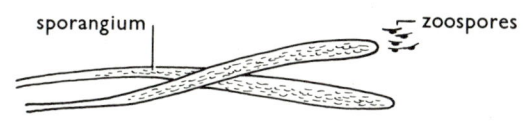

14.5 ASCOMYCOTA

Parasitic and saprophytic fungi, with septate hyphae, often found on wood.

14.5.1 Yeast (*Saccharomyces* **spp.**)

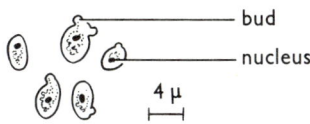

YEAST CELLS

In nature this fungus occurs on the skins of fruits, but owing to its importance in fermentation (see 21.2) it is cultivated for brewing and bread-making. It consists of minute cells which reproduce rapidly by budding. Examine a few cells on a slide, mounted in a drop of dilute sugar solution.

14.5.2 Strawberry Fungus (*Nectria cinnabarina*)

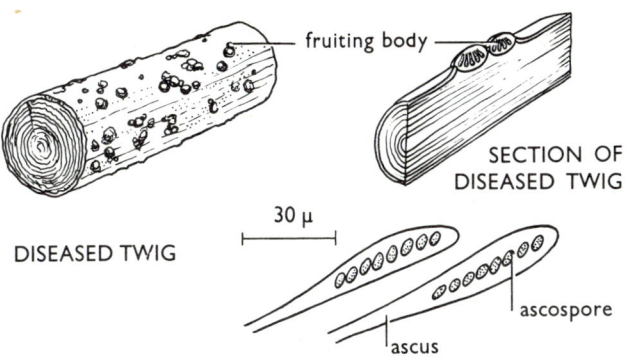

This is commonly found on twigs and fences, inside which its *hyphae* may be found. The fruiting bodies are bright red in colour. Pick a small portion of fruiting body off a twig and tease it on a slide. Mount in a drop of water and examine under low and high power. The fruiting body contains *asci*, flask-shaped structures each of which contains 8 *ascospores*.

14.6 BASIDIOMYCOTA

Parasitic and saprophytic fungi, with large, often brightly coloured, fruiting bodies.

14.6.1 The Mushroom (*Psalliota campestris*)

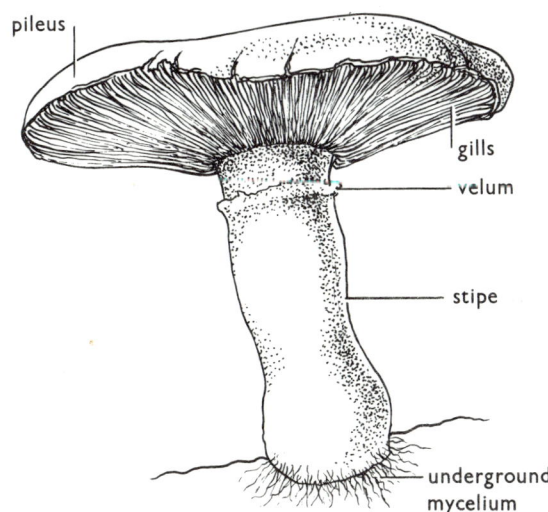

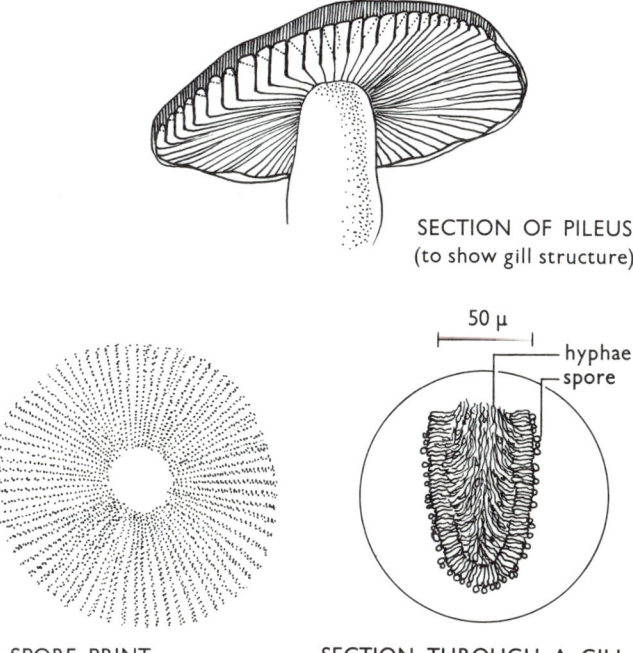

SECTION OF PILEUS
(to show gill structure)

SPORE PRINT   SECTION THROUGH A GILL

The *mycelium* extends through the ground, where it obtains food saprophytically from organic matter. When mature, the fruiting body develops rapidly, and in a few hours it appears above the soil. When fully grown it consists of a *pileus* under which there are many radial gills, each producing thousands of spores. The whole structure is made from tightly packed *hyphae*.

Place a mushroom *pileus* on a piece of white paper and leave it overnight. Spores ejected from the gill will be seen as a spore print on the paper. Examine prepared slides of transverse sections through the gills. Examine a piece of *mycelium* on a microscope slide; its *hyphae* are divided by septa into binucleate cells.

Simple green plants living in damp habitats. Reproduction occurs in a gamete-bearing generation, alternating with a sporophyte which is parasitic upon it.

15.1   The Common Moss (*Funaria hygrometrica*)

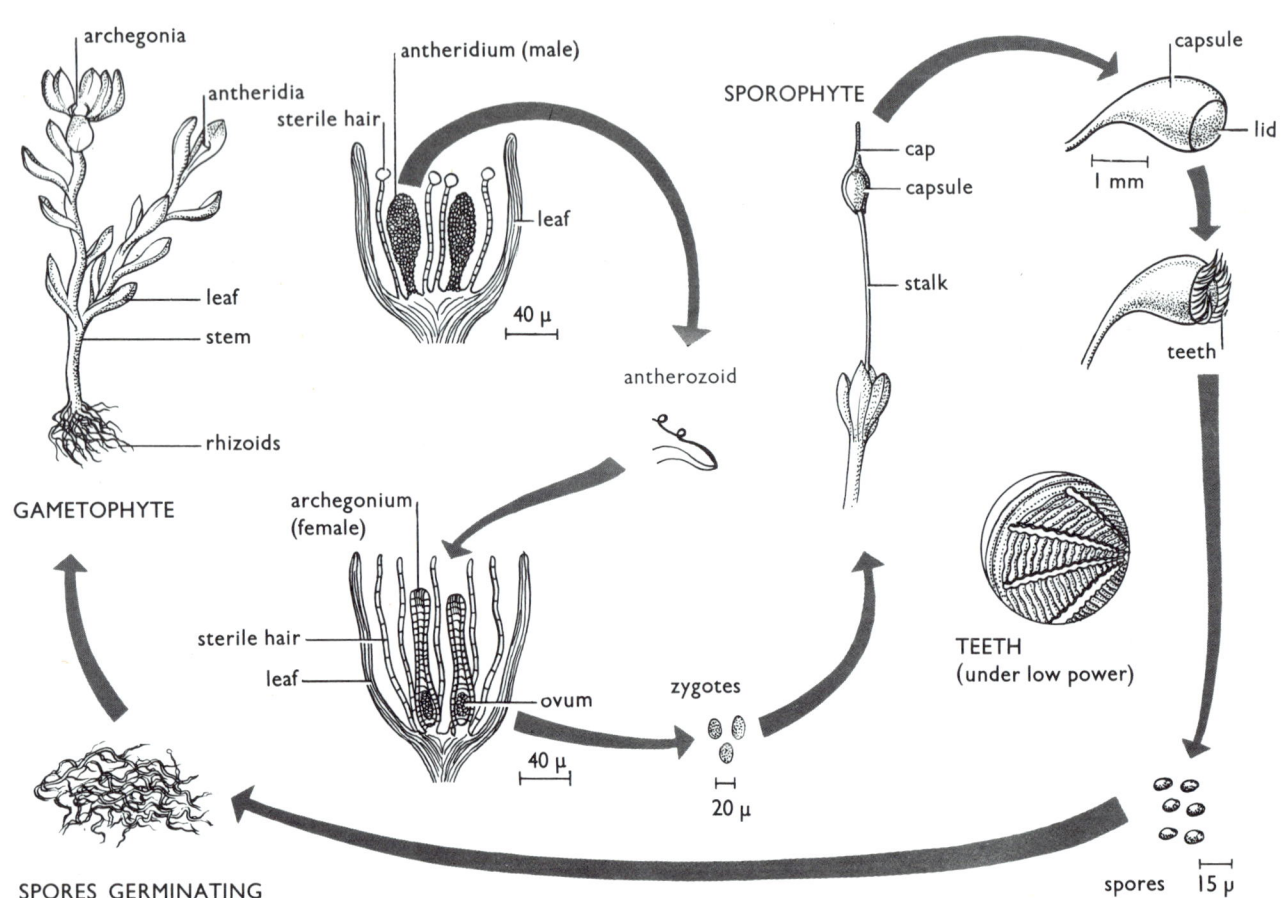

Habitat

It grows in tufts between bricks, stones, etc., and on damp soil.

Structure and nutrition

The gametophyte has a central stem, from which whorls of minute leaves emerge. They contain chlorophyll and make food by photosynthesis. *Rhizoids* anchor the plant to the soil.

Reproduction

The male organs are *antheridia*, club-shaped structures borne at the apex of a branch, protected by the terminal leaves and sterile hairs. Inside them *antherozoids* are produced (motile male gametes). When ripe they swim through the damp film of water covering the plant to the female organs, *archegonia*, similarly placed on the tip of another branch. An ovum develops at the base of each *archegonium*. The *antherozoids* are attracted chemically to the neck of the *archegonium*, and fertilization takes place when one swims down the canal and fuses with the ovum.

The fertilized zygote then develops as a parasite on the parent gametophyte and sends up a stalk on whose tip a spore capsule develops. This sporophyte remains attached to the gametophyte by a parasitic foot. When the capsule is ripe, a cap covering it drops off, and reveals a lid. This in turn splits away, and below it lie several radially arranged teeth. These move slightly as the humidity of the air varies, with the result that spores from the capsule are flicked out.

A spore germinates to produce a mass of green threads, from which a new gametophyte develops.

Method of examination

Examine growing plants with a hand lens and on a slide under the low power of the microscope. Distinguish between the gametophyte and sporophyte. Examine a ripe spore capsule and, if the teeth can be seen, breathe on them gently and observe their movements as the humidity of the air is changed by the breath.

Examine prepared slides of male and female reproductive organs and a vertical section through the capsule.

16 Pterophyta

Vascular plants, in which the sporophyte is well developed but the gametophyte is a simple prothallus.

16.1 The Male Fern (*Dryopteris filix-mas*)

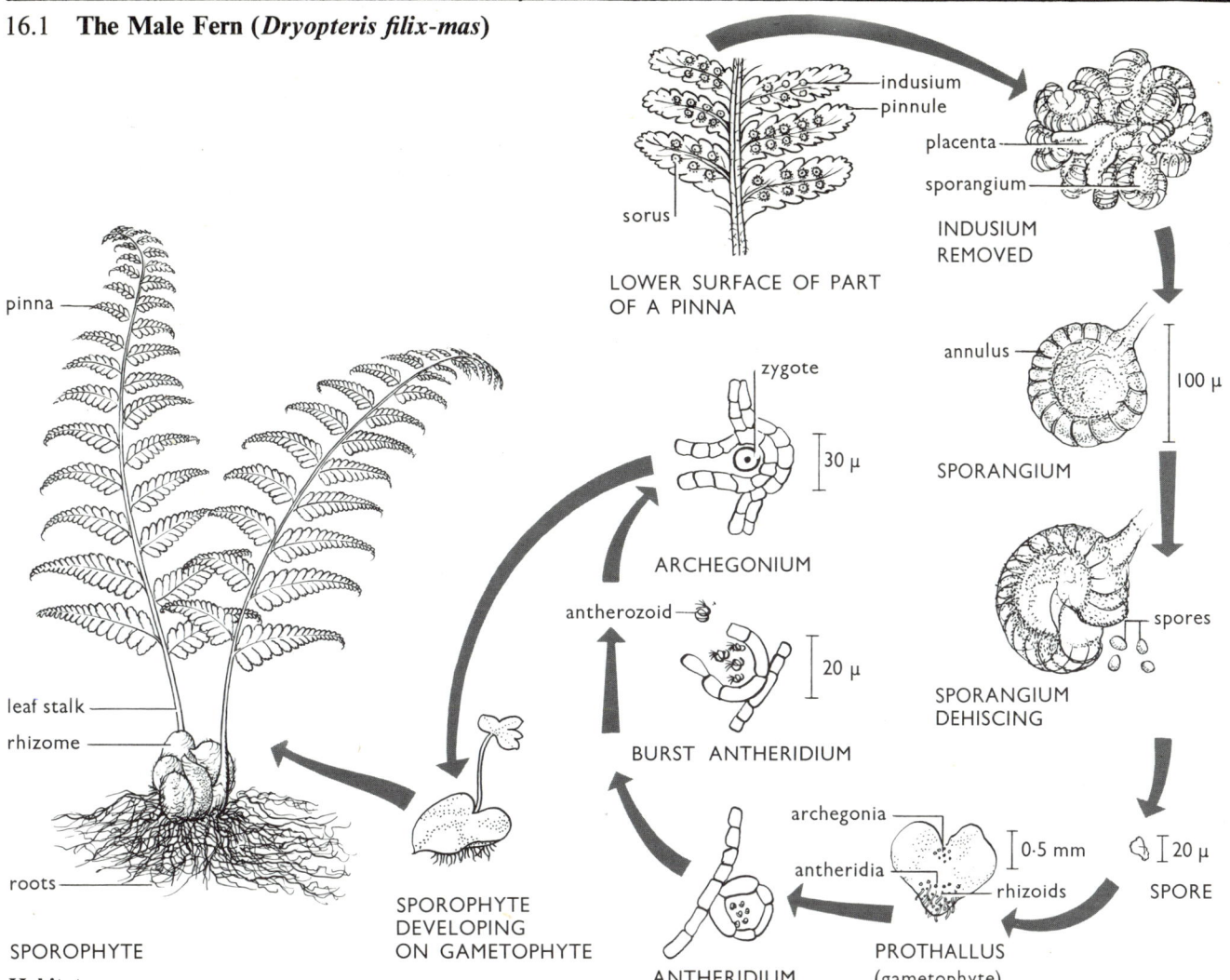

Habitat

In shady places in damp, slightly acid soil.

Structure

The sporophyte has a thick cylindrical underground *rhizome*, with a woody vascular strand in the centre of it. The tip of the *rhizome* is tilted towards the surface of the soil and from it the aerial leaves arise. The rest of the *rhizome* is covered with leaf bases, the remains of earlier leaves. Numerous roots occur on the lower surface anchoring it to the soil and absorbing water from it. The aerial leaves have a central stalk from which lateral leaflets, the *pinnae*, arise in considerable numbers. They are toothed structures sub-divided into *pinnules* containing chlorophyll. The plant makes its food by photosynthesis.

Reproduction

Underneath the fully formed *pinnae*, brown kidney-shaped *sori* occur. These consist of an *indusium*, an umbrella-shaped structure inside which are numerous *sporangia* radiating from a *placenta*. They are pear-shaped bodies with a prominent line of cells, the *annulus*, running round the circumference. When the *sporangium* is ripe it splits open and, by a sudden movement of the *annulus* cells, the spores are shot out.

The spores are carried by the wind, and if they alight on damp soil, they germinate to produce a small heart-shaped *prothallus* about 1 cm across. This is attached to the soil by

rhizoids from its lower surface. Also on the lower surface the reproductive organs develop; the males are simple *antheridia* inside which *antherozoids* are produced. The females are flask-shaped *archegonia*. Fertilization takes place when *antherozoids* emerge from the *antheridia* and swim to the ovum inside each *archegonium*. The fertilized zygote starts its life as a parasite on the gametophyte, but when it emerges above the soil it becomes photosynthetic, and develops into an independent sporophyte.

Method of examination

1. Examine a fully grown sporophyte, record the structure of the rhizome, its roots and the leaves.

2. Mount a piece of pinna dry on a slide, and observe the structure of the indusium and sporangia with the low power.

3. With a pair of needles, tease off a few sporangia on to a slide, mount in a drop of water and cover with a coverslip. Examine their structure, and the spores under low and high power.

4. Examine prepared slides of a transverse section through a sorus, showing indusium and sporangia.

5. Mount a gametophyte in a drop of water on a slide and cover with a coverslip. Examine its structure, and try to observe motile antherozoids. (See 23.3.3.)

6. Examine a prepared slide of a gametophyte.

17 Seed-bearing Plants

17.1 CONIFEROPHYTA

Arborescent plants, whose reproductive organs develop inside cones.

17.1.1 The Pine (*Pinus sylvestris*)

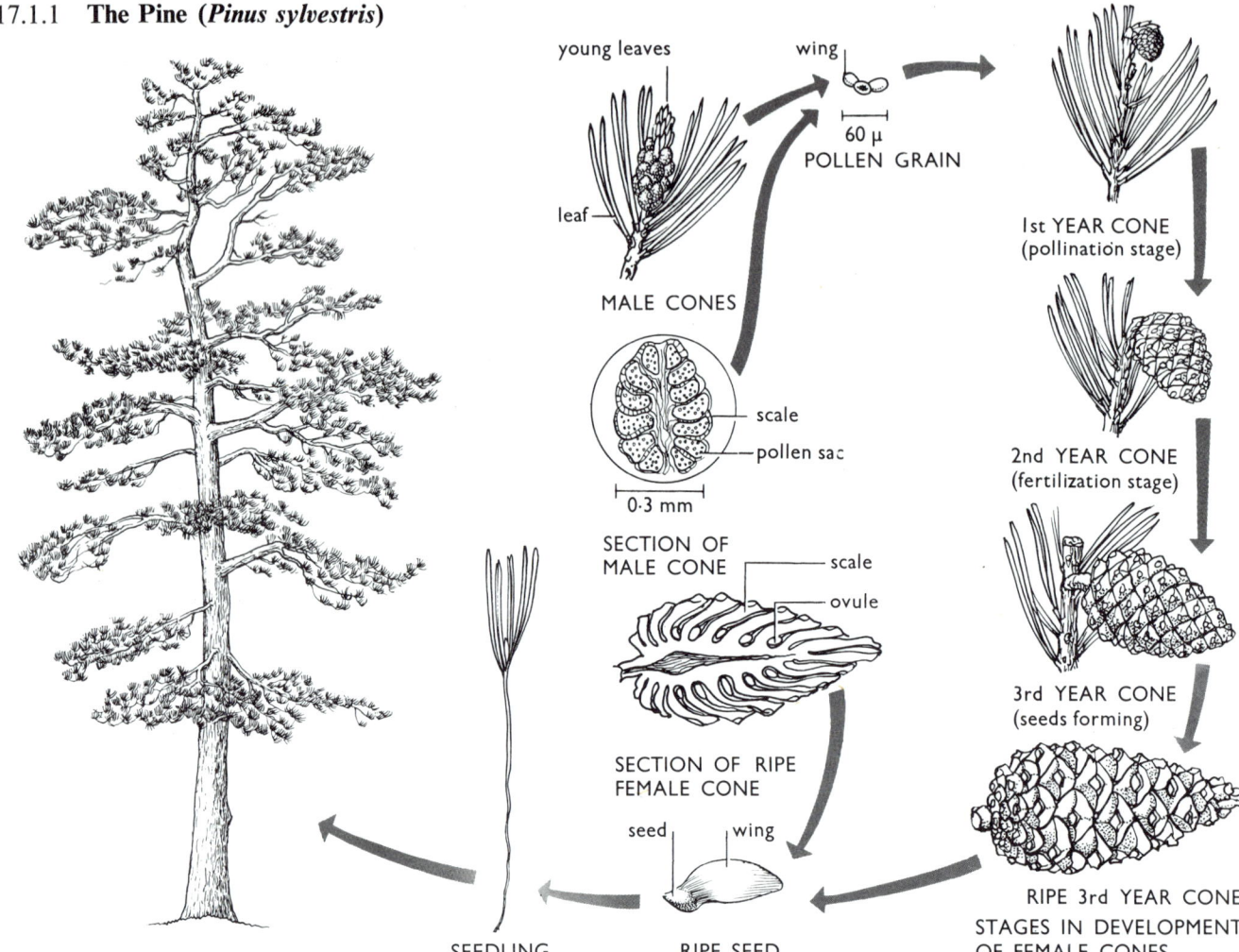

young leaves

leaf

MALE CONES

wing

60 μ
POLLEN GRAIN

scale

pollen sac

0·3 mm

SECTION OF MALE CONE

scale

ovule

SECTION OF RIPE FEMALE CONE

seed

wing

SEEDLING

RIPE SEED

1st YEAR CONE (pollination stage)

2nd YEAR CONE (fertilization stage)

3rd YEAR CONE (seeds forming)

RIPE 3rd YEAR CONE
STAGES IN DEVELOPMENT OF FEMALE CONES

Habitat

Its natural habitat is sandy soil or poor soils in mountainous districts.

Structure

A tree, which is fully grown in about 100 years. There is a straight trunk, from which the branches arise horizontally. The bark is thin and often exudes resin, a substance which protects the tree against the entrance of fungi. The wood is soft in texture and contains prominent *medullary rays* in which food is stored. The *xylem* is mostly made up of *fibres* and *tracheides*, the latter having *bordered pits*. The leaves are needle shaped, arranged in pairs and borne on short spurs emerging from the branches. They have a waxy epidermis, and the stomata are sunk in pits to prevent excessive water loss. The root system often has a prominent tap-root.

Reproduction

A mature tree produces male cones in clusters near the tips of its branches. These cones contain *pollen sacs*, inside which *pollen grains* are produced which ripen in early May. Each *pollen grain* has a pair of wings and is easily carried by the wind to the female cones. These are small green structures occurring singly near the tips of the branches, and they take one year to mature. *Pollen grains* alighting between the scales of one-year-old cones are protected between the scales, and fertilization of the ova takes place one year later, i.e. in the second year of the female cone's development. Two seeds develop at the base of each scale of the female cone, and they are ripe in the third year. A ripe cone is brown in colour and opens its scales on a dry day. The seeds, which are small, light and winged, are dispersed by the wind. They germinate in suitable surroundings and produce a small seedling with many cotyledons. During the first year the seedling grows little more than 12 cm.

Method of examination

1. Examine the terminal parts of branches, noting the structure of the leaves, male cones and female cones at different stages of growth.

2. Mount *pollen grains* obtained from male cones in a drop of water on a slide, and examine under low and high power.

3. Examine seeds with a hand lens.

4. Examine a section of a small branch with a hand lens, noting the annual rings of growth, *medullary rays*, phloem and *bark*.

5. Examine prepared slides of sections through a stem and leaf.

17.2.1 General Structure

Flowering plants typically have four main organs. (*a*) The flower concerned in reproduction. (*b*) The stem which supports the flowers and leaves and transports food and water. It sometimes stores food in its tissues. (*c*) The leaves borne on *petioles* attached to the stem at regions called *nodes*. In the angle between the petiole and the stem a bud may be found, said to be *axillary* in position. (*d*) The root system anchors the plant in the soil, from which it absorbs moisture through a system of root hairs. Many plants store food in their roots.

Examine a specimen of a buttercup and identify the main organs and their arrangement.

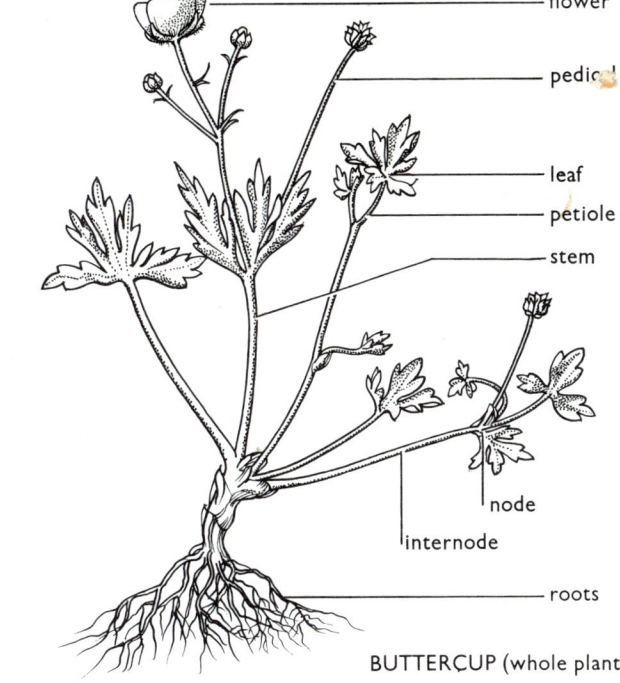

BUTTERCUP (whole plant)

LOW POWER DIAGRAM

The Root

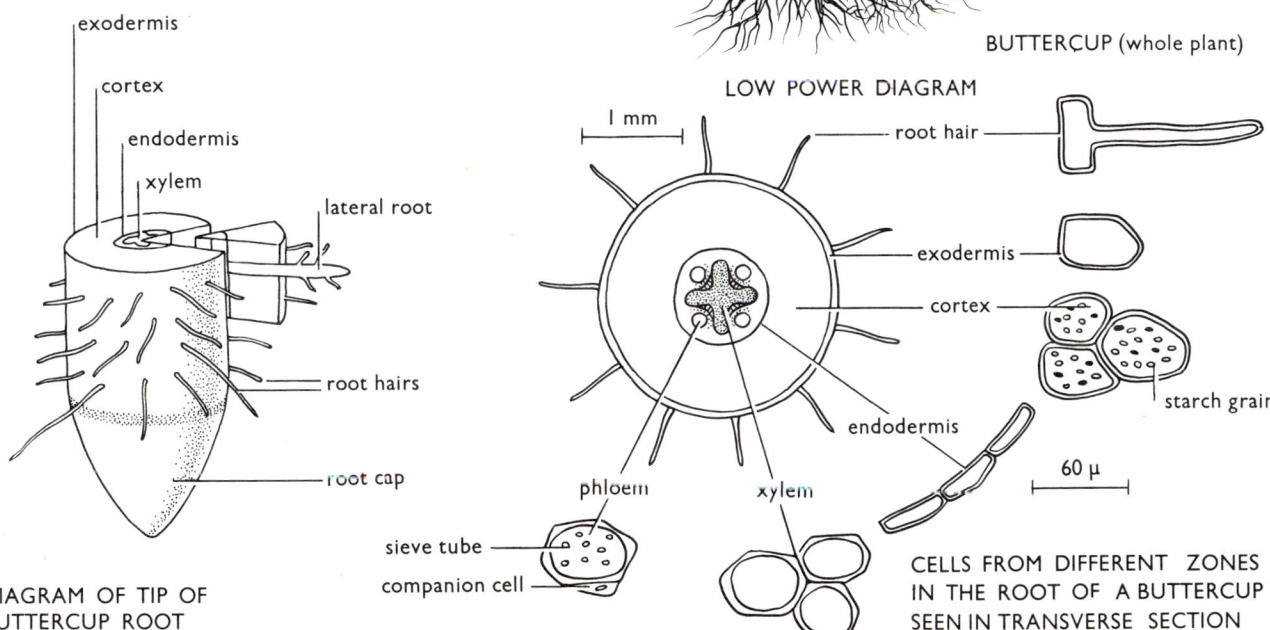

DIAGRAM OF TIP OF
BUTTERCUP ROOT

CELLS FROM DIFFERENT ZONES
IN THE ROOT OF A BUTTERCUP
SEEN IN TRANSVERSE SECTION

There are strong roots from which numerous lateral roots emerge from the inner tissues. A transverse section through a root shows the arrangement of the tissues. From the outside inwards they are:

ROOT HAIRS: a ring of cells on the outside of the root a few millimetres behind its tip. They absorb water from the soil.

EXODERMIS: a layer of cells particularly well developed in regions away from the root hairs. They are waterproof.

CORTEX: large vacuolated cells which store starch.

ENDODERMIS: a ring of cells surrounding the central vascular cylinder.

XYLEM: cells with woody walls arranged in the shape of a four-rayed star. They conduct water from the roots up to the stem.

PHLOEM: thin-walled cells, tubular in shape. They conduct food solutions from the stem into the cortex for storage in the form of starch.

At the root tip there is a cap of slimy cells, which protect it as it grows through the soil.

Method of examination

Examine whole roots with a hand lens and observe lateral roots and root hairs. Mount thin sections of buttercup root on a slide and examine under low and high power. Mount one section in a drop of aniline sulphate which stains xylem cell walls yellow. Mount another section in a drop of dilute iodine to reveal the distribution of starch grains.

Examine a prepared slide of a transverse section.

The Stem

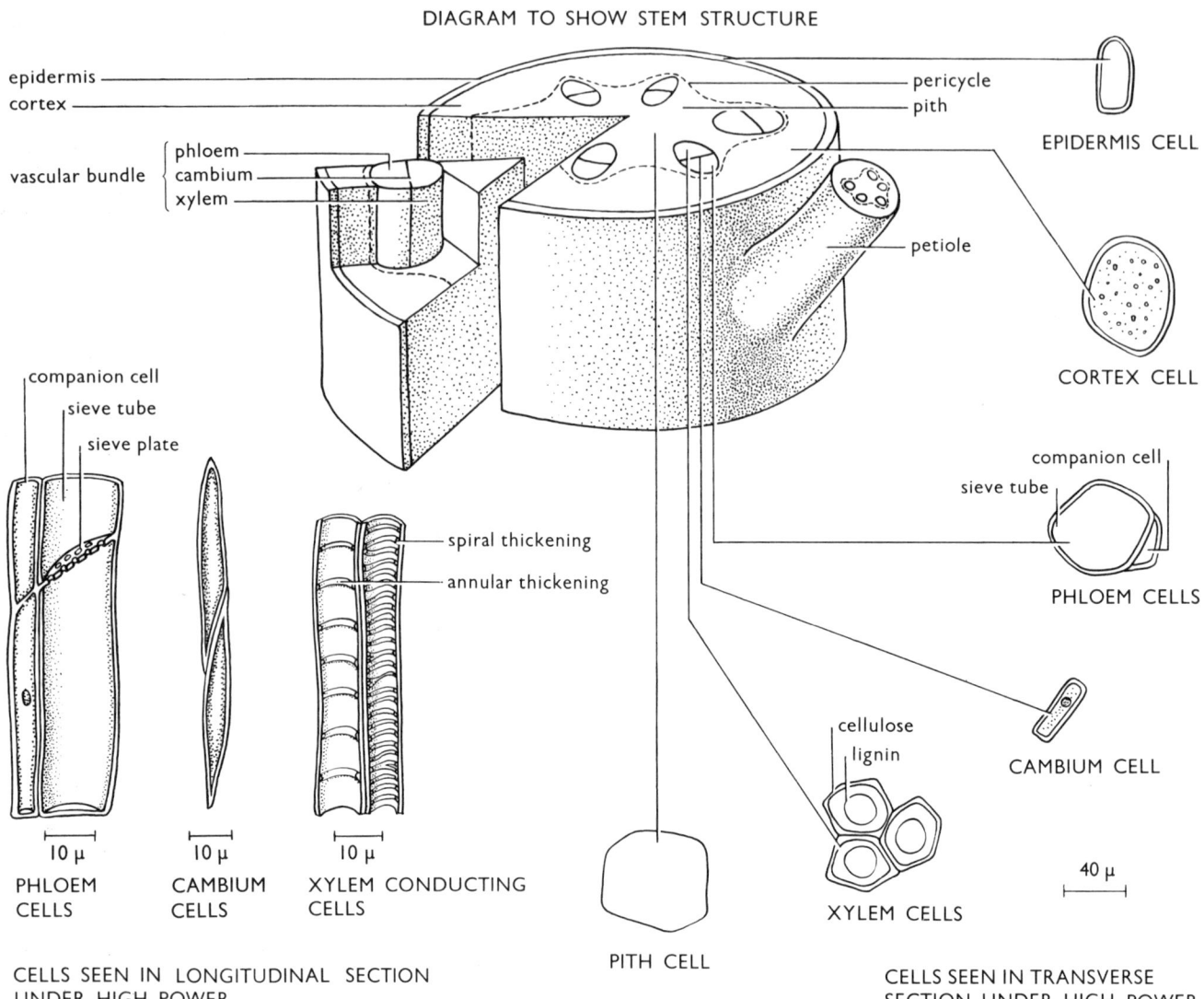

DIAGRAM TO SHOW STEM STRUCTURE

PHLOEM CELLS / CAMBIUM CELLS / XYLEM CONDUCTING CELLS

CELLS SEEN IN LONGITUDINAL SECTION UNDER HIGH POWER

PITH CELL

EPIDERMIS CELL

CORTEX CELL

PHLOEM CELLS

CAMBIUM CELL

XYLEM CELLS

CELLS SEEN IN TRANSVERSE SECTION UNDER HIGH POWER

The distribution of the tissues from the outside of the stem inwards, is as follows:

EPIDERMIS: A single layer of cells, whose outside walls are covered with wax making the stem waterproof. Occasionally pores, the *stomata*, occur in it.

CORTEX: A rind of compact vacuolated cells. The outermost may contain chloroplasts, and they have cell walls thickened at the corners. The inner cortex cells have thin walls of even texture; this is *parenchyma* tissue.

PERICYCLE: A single layer of cells, separating the *cortex* from the *pith* and *vascular* tissues.

VASCULAR BUNDLES: These lie within the *Pericycle* and contain three types of tissue.

(a) *Phloem*. Tissues in which food solutions travel up or down the stem in *sieve tubes*, so called because at intervals there are perforated plates placed transversely across them. Each *sieve tube* has a *companion cell* (with a prominent nucleus) lying alongside it. Some *parenchyma* tissue may also be observed among the sieve tubes.

(b) *Cambium*. A thin layer of rectangular cells which remain active throughout the life of the stem, and form new

phloem tissues outwards, and new *xylem* tissues inwards.

(c) *Xylem*. Cells, whose *cellulose* walls are thickened with *lignin*, giving them their woody character. There are three types: *conducting cells* in which water ascends the stem; *fibres*; *xylem parenchyma*.

PITH: Large *parenchymatous* cells in the centre of the stem. In old stems they frequently die, leaving the centre hollow.

MEDULLARY RAYS: A region of tissue lying radially in position between the vascular bundles.

Method of examination

Examine a whole stem, and identify the *nodes, internodes, axillary buds* and *petioles*. Examine thin transverse sections; mount one in aniline sulphate which stains woody tissues yellow, and another in iodine to observe the distribution of starch grains. Examine a prepared slide of transverse and longitudinal sections.

In drawing stems, make a simple plan of the arrangement of the tissues as seen under the low power. Then draw one or two cells from each of the main regions as seen under the high power. Do not attempt to draw several cells of the same kind.

Secondary Thickening

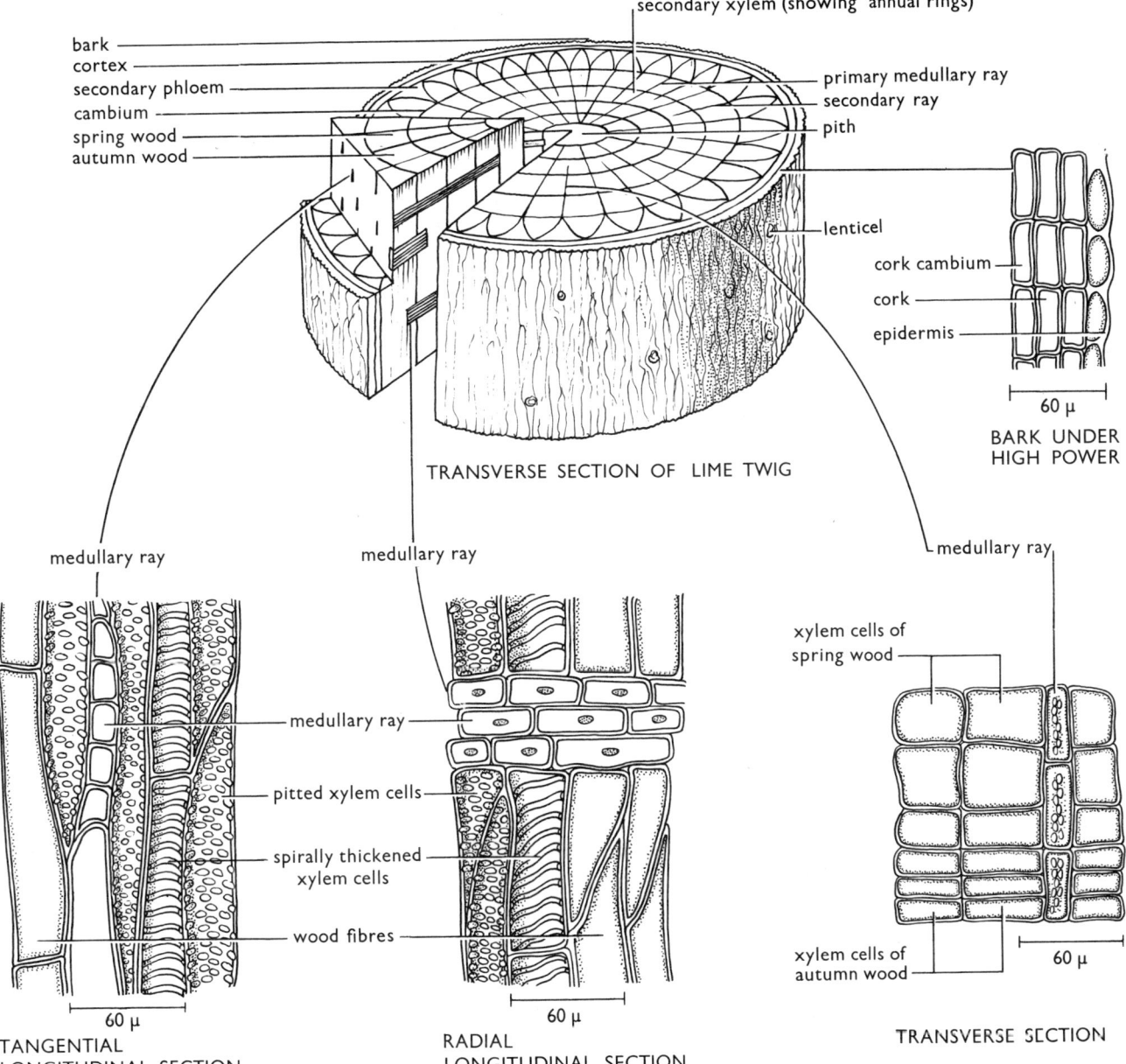

bark
cortex
secondary phloem
cambium
spring wood
autumn wood

secondary xylem (showing annual rings)

primary medullary ray
secondary ray
pith

lenticel

TRANSVERSE SECTION OF LIME TWIG

cork cambium
cork
epidermis

60 μ

BARK UNDER
HIGH POWER

medullary ray

medullary ray

medullary ray

medullary ray

pitted xylem cells

spirally thickened
xylem cells

wood fibres

60 μ

TANGENTIAL
LONGITUDINAL SECTION

60 μ

RADIAL
LONGITUDINAL SECTION

xylem cells of
spring wood

xylem cells of
autumn wood

60 μ

TRANSVERSE SECTION

In shrubs and trees where growth continues from year to year, the stem increases in size by a process of secondary thickening. The *cambium* becomes active and forms a complete cylinder round the stem between the cortex and pith. It produces a secondary growth of *xylem* and, as it is more active in the spring than in autumn, this results in the production of *spring wood*—large cells with small deposits of lignin, and *autumn wood*—smaller cells with thick lignified walls. This gives the appearance of rings of annual growth. Running radially through the xylem are cellulose-walled cells of the *medullary rays*. In these food is stored, and they give lines of weakness in the xylem tissues. The phloem cells are pushed outwards by the *cambium*, and *secondary phloem* tissues are somewhat squashed and occupy less space than the xylem cells.

The *epidermis* is replaced by a layer of *cork*, produced by a *cork cambium* which develops in the outer regions of the cortex. The stomata are replaced by larger breathing pores— the *lenticels*.

In very old trees, the central xylem cells cease to conduct water and become the darkly stained *heart wood*. The outer xylem which continues to conduct water is known as *sap wood*.

Method of examination

Examine thin sections of three-year-old twigs of Lime (*Tilia europea*), mounted in aniline sulphate. In the phloem tissue there are alternate bands of sieve tubes and lignified fibres. Examine prepared slides of Transverse Sections, Radial Longitudinal, and Tangential Longitudinal Sections of woody stems.

The Leaf

GRASS CLOVER SYCAMORE COMMON ASH

NASTURTIUM BEECH OAK HORSE CHESTNUT

VARIETIES IN FORM

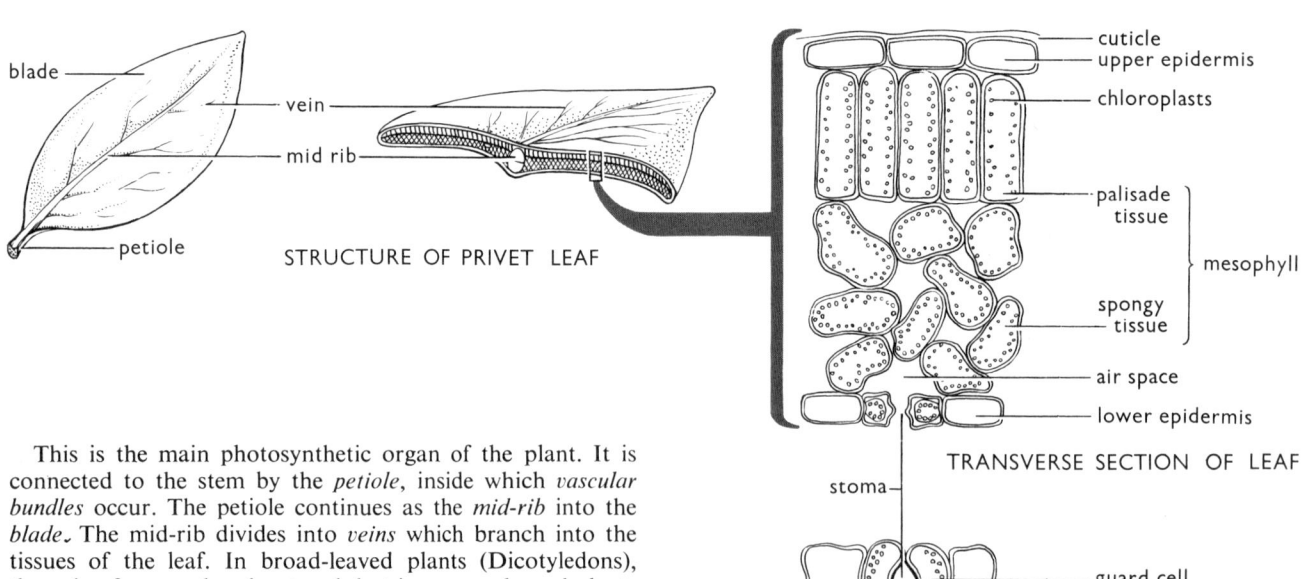

STRUCTURE OF PRIVET LEAF

blade — vein — mid rib — petiole

cuticle
upper epidermis
chloroplasts
palisade tissue
mesophyll
spongy tissue
air space
lower epidermis

TRANSVERSE SECTION OF LEAF

stoma

guard cell
chloroplasts

SURFACE VIEW OF STOMA

This is the main photosynthetic organ of the plant. It is connected to the stem by the *petiole*, inside which *vascular bundles* occur. The petiole continues as the *mid-rib* into the *blade*. The mid-rib divides into *veins* which branch into the tissues of the leaf. In broad-leaved plants (Dicotyledons), the veins form a closed network but in narrow-leaved plants (Monocotyledons) the veins are open at their ends.

In transverse section the main tissues are:

UPPER EPIDERMIS (Adaxial surface). Close-fitting rectangular cells, with a waxy cuticle covering the outer surface.

MESOPHYLL. (*a*) *Palisade cells*—column shaped and containing *chloroplasts*. (*b*) *Spongy tissue*—loose-fitting cells with air spaces between them. The vascular bundles (veins) run through the *mesophyll*, the xylem towards the adaxial surface.

LOWER EPIDERMIS (Abaxial surface). Close-fitting rectangular cells, with a waxy cuticle covering the outer surface. Stomata occur in it, each one being an opening surrounded by two sausage-shaped cells—the *guard cells*.

Method of examination

1. Examine leaves from different plants, identify them, and record their shapes.

2. Examine a transverse section of a Privet leaf mounted in a drop of water. Draw a low power diagram and, under the high power, a few cells.

3. Examine the surface of leaves under the low power and study the arrangement of the stomata.

4. Tear thin strips of epidermis from an Iris leaf, and examine its stomata in detail. (See 19.2.)

The Flower

The Buttercup (*Ranunculus* spp.)

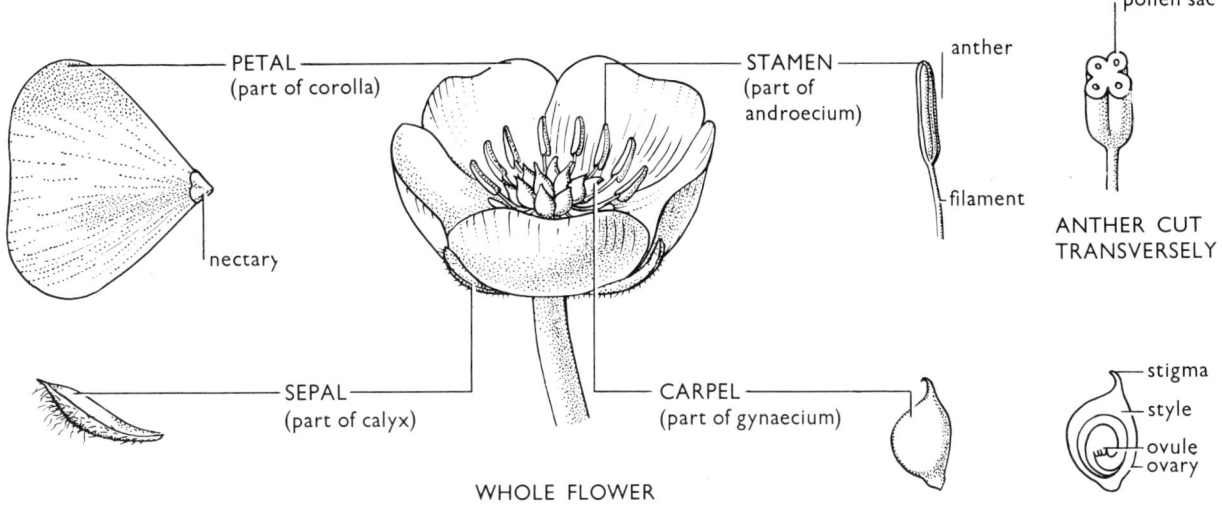

WHOLE FLOWER

ANTHER CUT TRANSVERSELY

CARPEL CUT LONGITUDINALLY

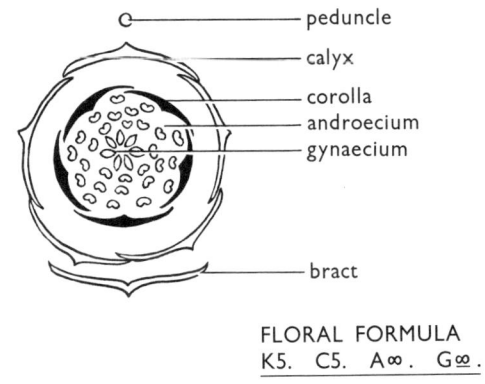

FLORAL FORMULA
K5. C5. A∞. G∞.

FLORAL DIAGRAM OF A FLOWER

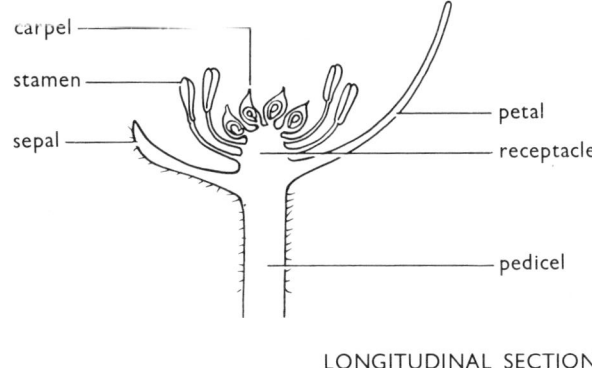

LONGITUDINAL SECTION OF A FLOWER

The flower is borne on a short lateral stem—the *pedicel*, which arises in the bud stage within a protective leaf—the *bract*. Bracts are not visible in the fully developed flower. The part of the flower nearest the main axis is posterior, the part away from the axis is anterior, and the bract occupies this position. The floral parts are arranged in four concentric circles or *whorls*, from the *receptacle*, which is the swollen tip of the *pedicel*. Starting from the outside and working inwards they are:

The calyx consisting of five green sepals.

The corolla consisting of five petals alternating between the sepals. At the base of each petal there is a nectary.

The androecium. There is an infinite number of *stamens* each consisting of a *filament* supporting an *anther*, inside which pollen grains develop.

The gynaecium. An infinite number of *carpels* within the stamens. Each one consists of an *ovary* containing one *ovule*,

and a short *style* surmounted by a *stigma* placed above the ovary.

The structure of the flower can be represented by a *floral diagram* showing the distribution in plan view of the four parts, and by a longitudinal slice through it. The accurate numbers of parts in each whorl are recorded by a *floral formula*.

Method of examination

1. From a fresh specimen remove a sepal, petal, stamen and carpel. Use a hand lens to examine, and draw each part. Mount a stamen on a slide in a drop of water and examine the anther and pollen grains under low and high power.

2. Examine a carpel on a slide in a drop of water under the low power.

3. Cut a longitudinal section through a whole flower with a sharp knife and record the arrangement of the parts.

4. Make a floral diagram and record the floral formula.

69

The Dandelion (*Taraxacum officinale*)

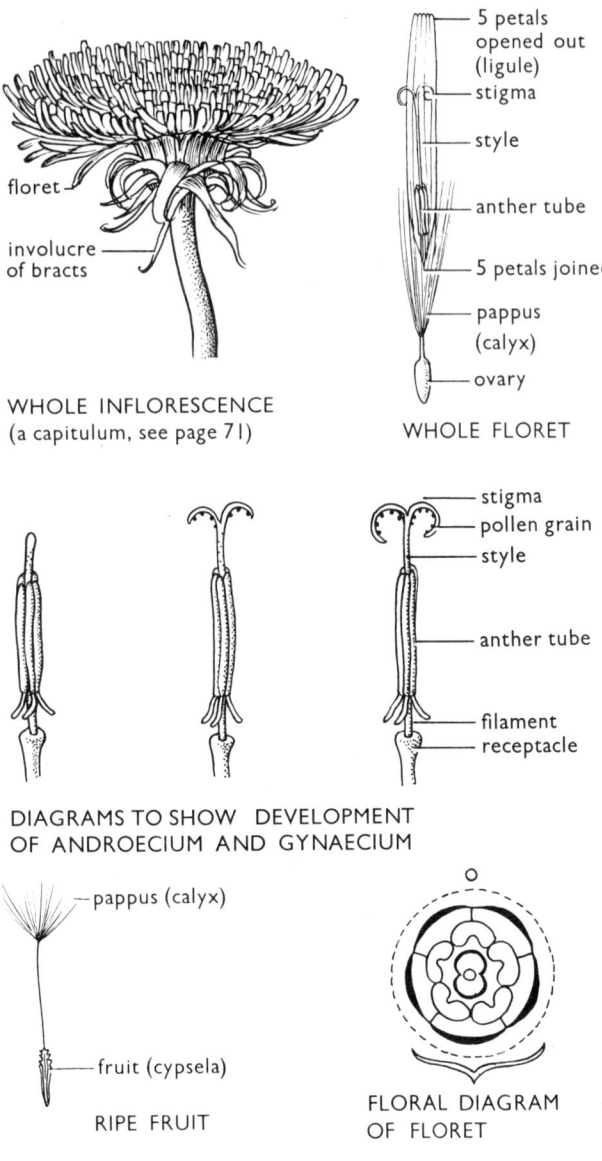

WHOLE INFLORESCENCE
(a capitulum, see page 71)

WHOLE FLORET

DIAGRAMS TO SHOW DEVELOPMENT
OF ANDROECIUM AND GYNAECIUM

RIPE FRUIT

FLORAL DIAGRAM
OF FLORET

FLORAL FORMULA K∞. C(5). A(5). G$\overline{(2)}$.

This is a composite flower, in which numerous *florets* are borne on the *capitulum*, an enlarged receptacle. The *capitulum* is protected by an *involucre* of green *bracts*. Each *floret* has the usual four *whorls* of floral parts. The *calyx* has hair-like processes which develop into the *pappus* of the fruit. The *corolla* has five petals joined together at the base, but opening out at the top into a flattened *ligule*. Within the *corolla tube* are five stamens joined together by their anthers. The gynaecium consists of two fused carpels containing one ovule. The style is a cylindrical structure which grows up through the stamenal tube opening into a two-sectioned stigma.

Method of examination

Examine the arrangement of the florets in the inflorescence with a hand lens. Remove one floret, place it on a white tile and examine it with a hand lens. Mount a floret dry on a slide and examine its structure under the low power. Record the floral diagram and formula.

The Sweet Pea (*Lathyrus* spp.)

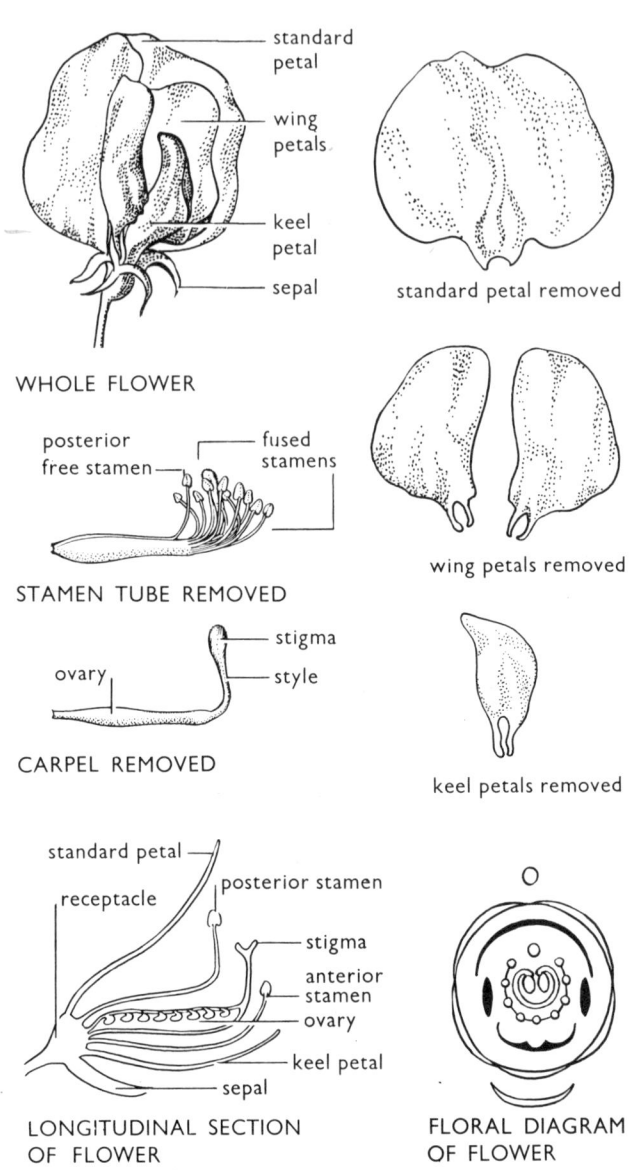

WHOLE FLOWER

standard petal removed

STAMEN TUBE REMOVED

wing petals removed

CARPEL REMOVED

keel petals removed

LONGITUDINAL SECTION
OF FLOWER

FLORAL DIAGRAM
OF FLOWER

FLORAL FORMULA K(5). C3 + (2). A(9) +1. G$\underline{1}$.

Two or 3 flowers are often found on 1 *pedicel*. The calyx has 5 *sepals*, fused by their bases. The corolla has 5 *petals*—The '*standard*', the posterior petal. The '*wings*', 2 lateral petals. The '*keel*', 2 anterior petals fused together and protecting the androecium. This has a tube of 9 stamens joined together, and 1 posterior stamen free from the others. The gynaecium has 1 carpel containing many ovules, the style and stigma curve upwards within the keel.

Method of examination

Examine whole flowers with a hand lens. Remove the sepals and petals one by one and draw them. Observe and record the structure of the androecium and gynaecium. Examine pollen grains on a slide under the high power of the microscope.

Cut a longitudinal section through the flower and make a drawing of the section. Observe the arrangement of the ovules in the ovary.

Record the floral diagram and floral formula.

Variations in Floral Structure

(a) Inflorescences

Peduncle. The main stalk of an inflorescence.
Pedicel. The main stalk of a flower.
Bract. The scale leaf, in whose axil the pedicel arises.

RACEMOSE INFLORESCENCES (main stalk does not end in a
flower, and continues growth)

CYMOSE INFLORESCENCES (the growth of the main stalk
is stopped by a terminal flower)

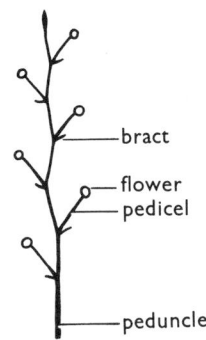

UMBEL, e.g. cowslip

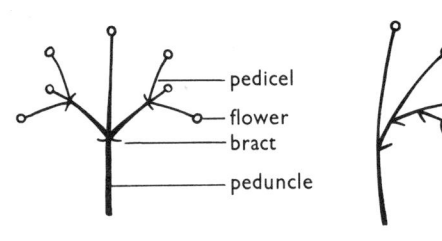

DICHASIUM, e.g. chickweed

MONOCHASIUM,
e.g. forget-me-not

RACEME, e.g. wallflower

CAPITULUM, e.g. dandelion

(b) Receptacles

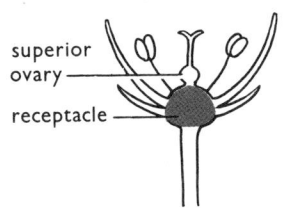

HYPOGYNOUS FLOWER, e.g. buttercup

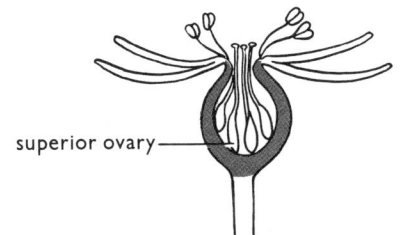

PERIGYNOUS FLOWER, e.g. rose

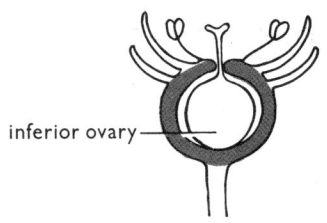

EPIGYNOUS FLOWER, e.g. apple

(c) Ovaries

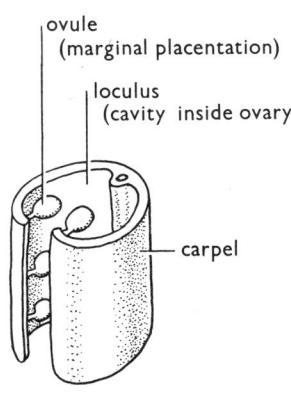

MONOCARPELLARY OVARY
e.g. pea

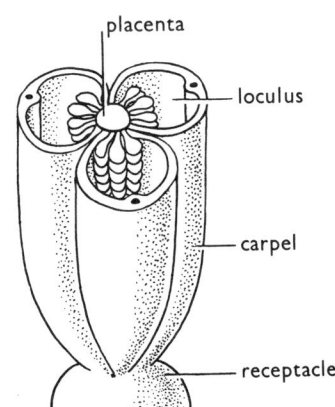

TRICARPELLARY OVARY
(ovules with axile placentation,
e.g. bluebell)

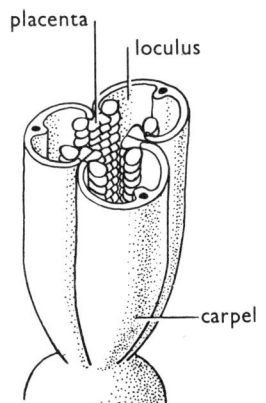

TRICARPELLARY OVARY
(ovules with parietal
placentation, e.g. viola)

71

17.2.2 POLLINATION MECHANISMS

(a) Monoecious
Unisexual flowers on the same plant.

(b) Dioecious
Unisexual flowers on different plants.

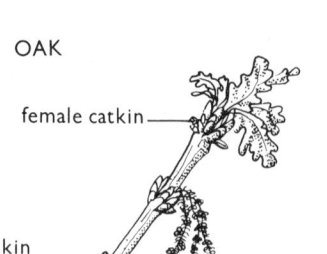

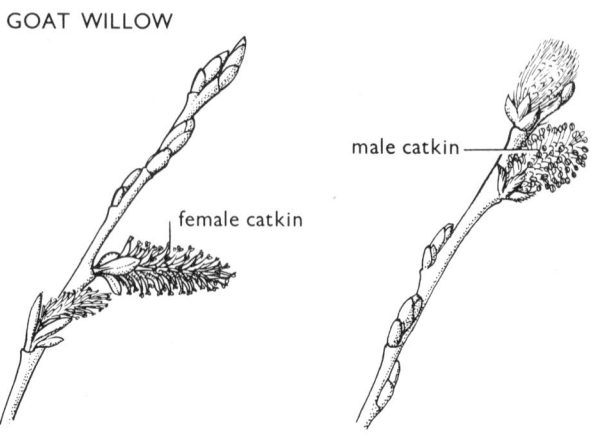

(c) Hermaphrodite
Male and female organs in the same flower.

COMPOSITAE (protandry. Anthers ripen before the stigmas)

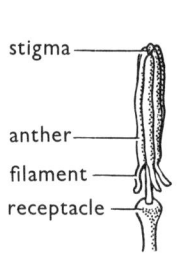

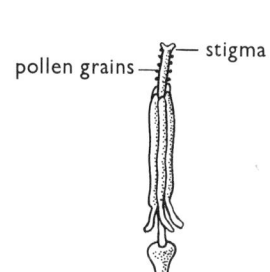

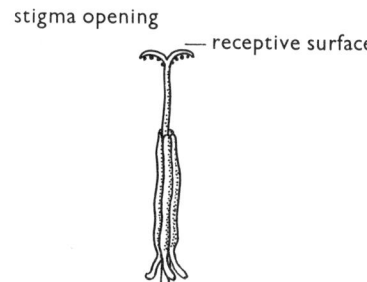

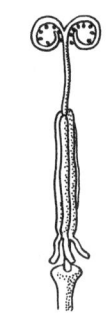

1. stigma growing through anther tube

2. pollen grains carried out on lower surface of stigma

3. pollen grains dispersed to other florets by insects

4. stigma self fertilizes itself with any pollen grains remaining on it

N.B. In the Dandelion, the seed usually develops without fertilization.

PLANTAIN (protogyny. Stigmas ripen before the anthers)

PRIMROSE (heterostyly. Flowers in two forms with different length of style)

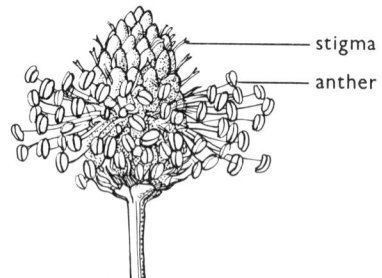

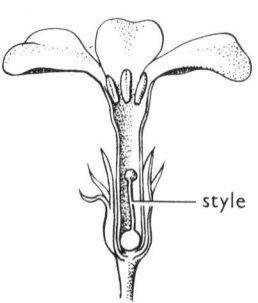

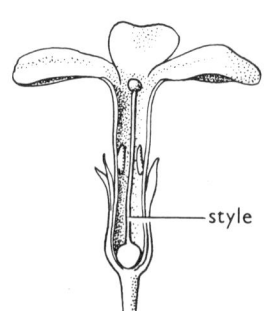

THRUM EYED (short style)　　　　PIN EYED (long style)

17.2.3 FRUITS AND SEEDS

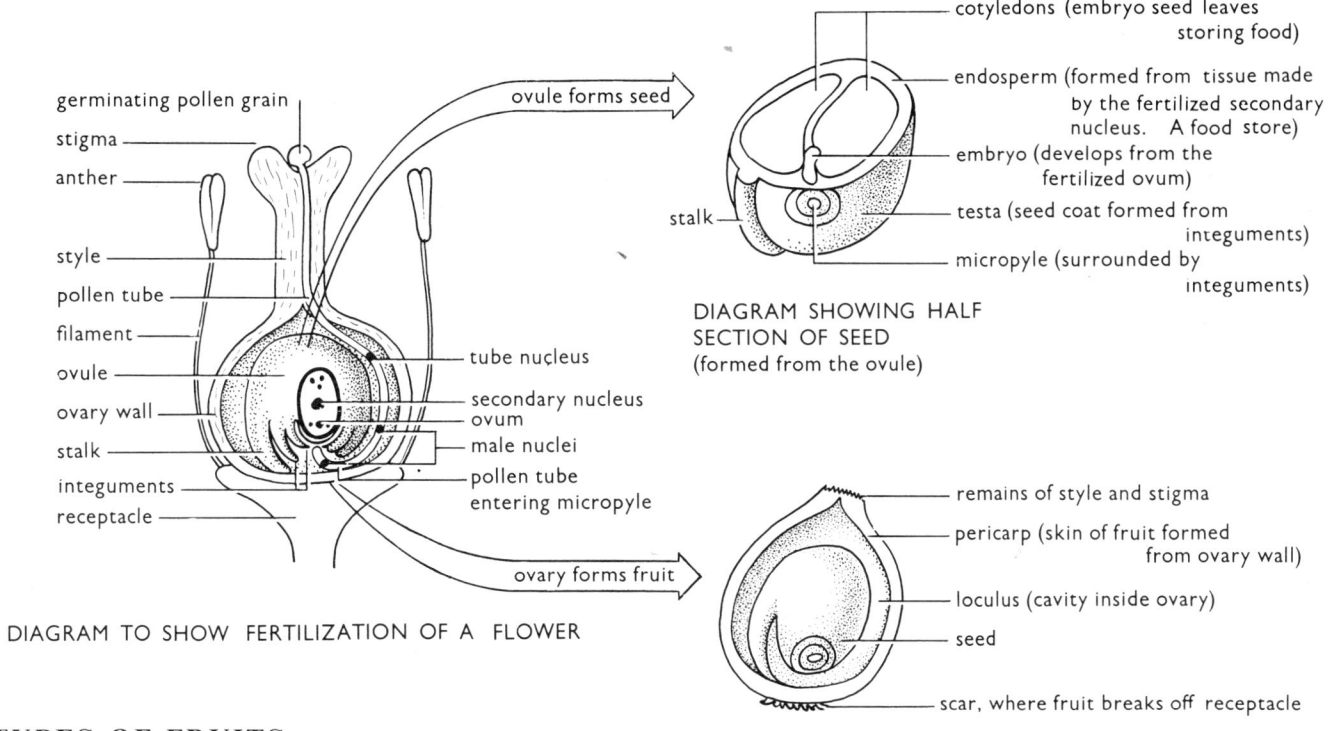

germinating pollen grain
stigma
anther
style
pollen tube
filament
ovule
ovary wall
stalk
integuments
receptacle

tube nucleus
secondary nucleus
ovum
male nuclei
pollen tube entering micropyle

ovule forms seed

ovary forms fruit

DIAGRAM TO SHOW FERTILIZATION OF A FLOWER

cotyledons (embryo seed leaves storing food)
endosperm (formed from tissue made by the fertilized secondary nucleus. A food store)
embryo (develops from the fertilized ovum)
testa (seed coat formed from integuments)
micropyle (surrounded by integuments)
stalk

DIAGRAM SHOWING HALF SECTION OF SEED
(formed from the ovule)

remains of style and stigma
pericarp (skin of fruit formed from ovary wall)
loculus (cavity inside ovary)
seed
scar, where fruit breaks off receptacle

DIAGRAM SHOWING VERTICAL SECTION OF FRUIT
(formed from ovary and ovule)

N.B. A seed has one scar where it breaks off its stalk. A fruit has two scars

TYPES OF FRUITS
Simple fruits (made from one flower)

DRY INDEHISCENT. Single-seeded fruits, in which the pericarp does not split to release the seed.

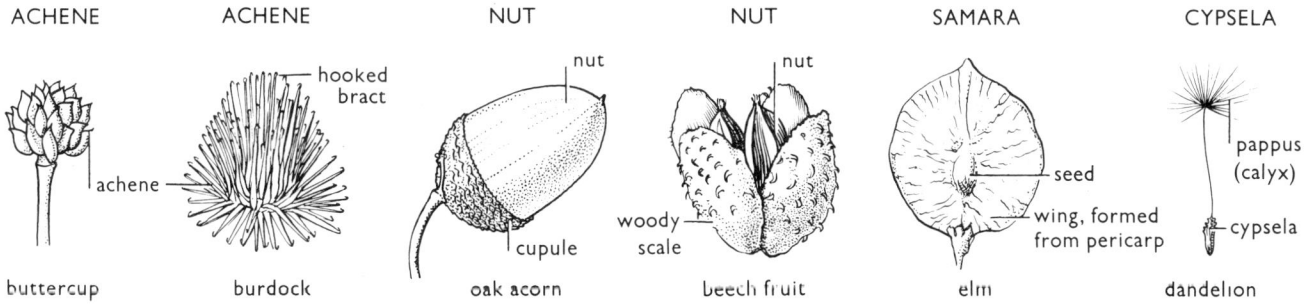

ACHENE

achene

buttercup

ACHENE

hooked bract

burdock

NUT

nut

cupule

oak acorn

NUT

nut

woody scale

beech fruit

SAMARA

seed

wing, formed from pericarp

elm

CYPSELA

pappus (calyx)

cypsela

dandelion

WINGED. The wall of the pericarp develops into a flattened wing, for dispersal by the wind.

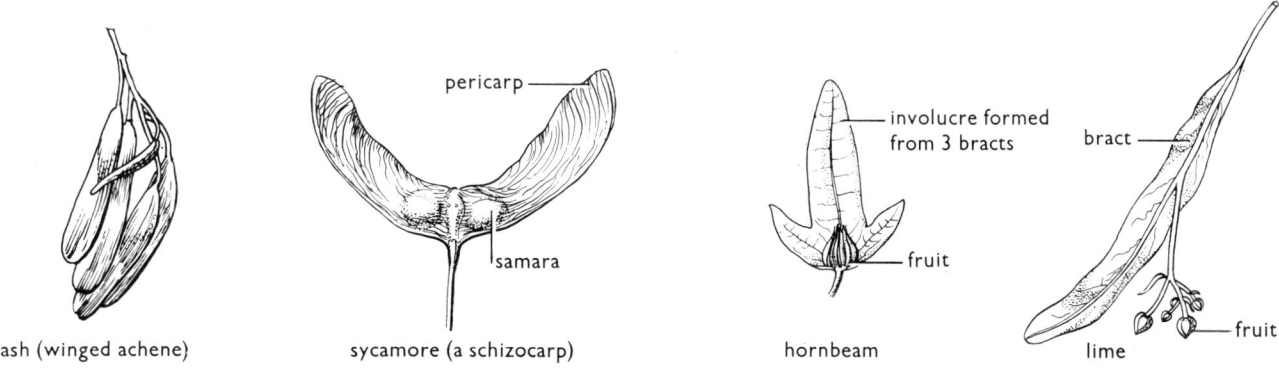

ash (winged achene)

pericarp

samara

sycamore (a schizocarp)

involucre formed from 3 bracts

fruit

hornbeam

bract

fruit

lime

DRY DEHISCENT. The seeds are released by the splitting of the pericarp.

FOLLICLE

paeony

LEGUME

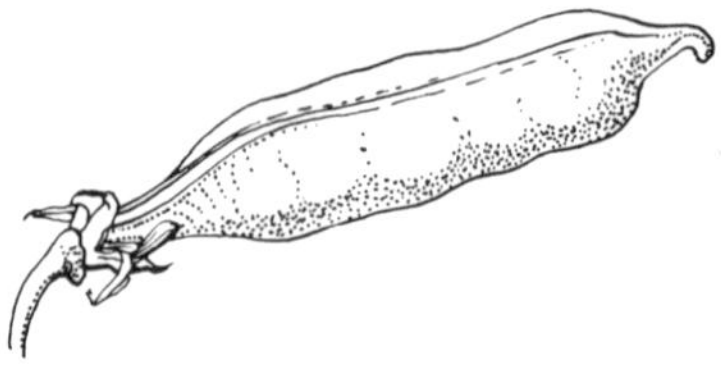

pea

CAPSULE

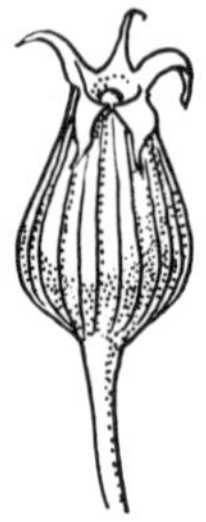

campion

CAPSULE

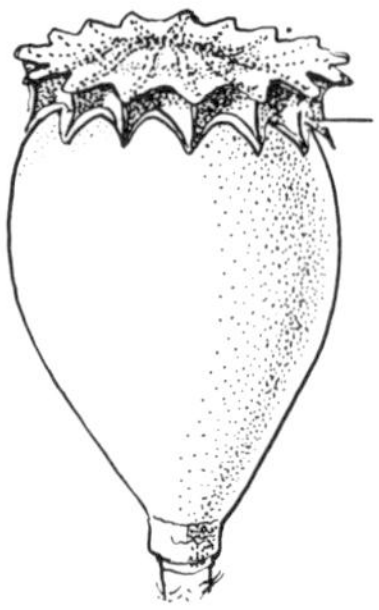

pore

poppy

poppy capsule in section

SILIQUA

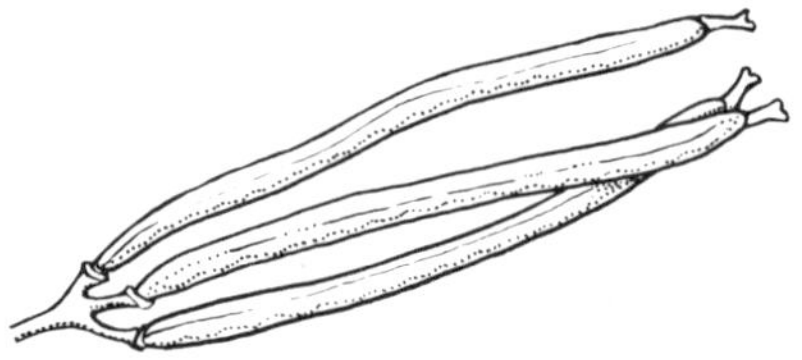

wallflower

SILICULA

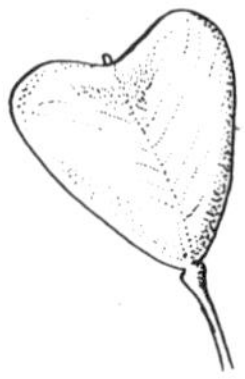

shepherd's purse

SUCCULENT. Part of the pericarp becomes fleshy.

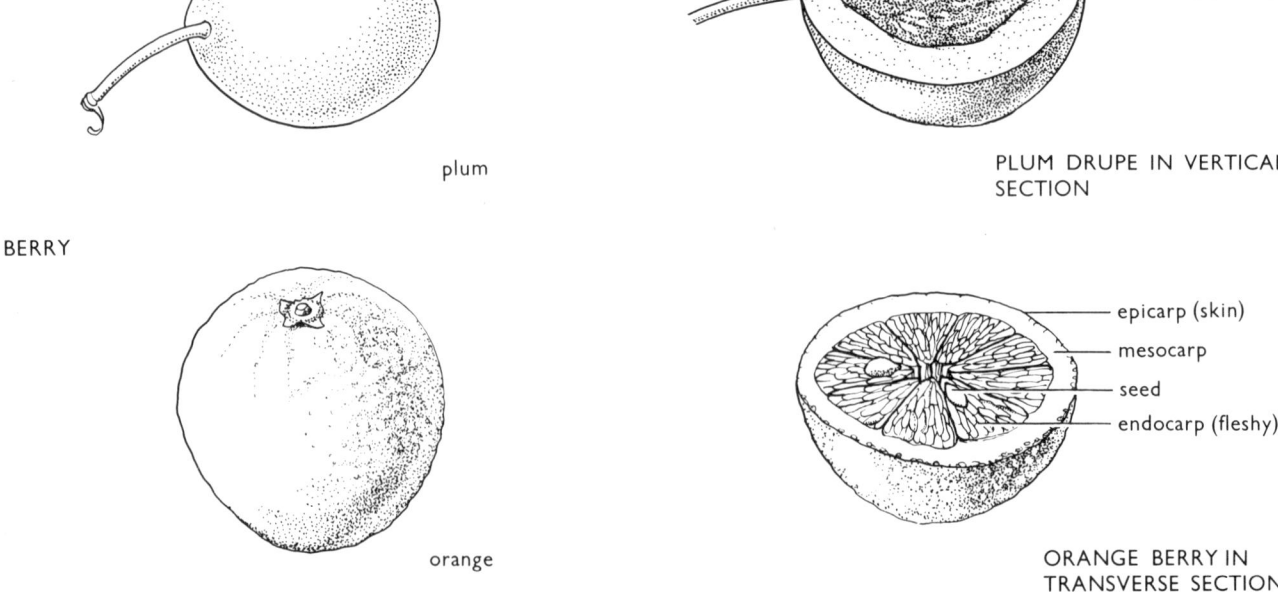

DRUPE

plum

epicarp (skin)

mesocarp (fleshy)

endocarp (stone)

PLUM DRUPE IN VERTICAL SECTION

BERRY

orange

epicarp (skin)

mesocarp

seed

endocarp (fleshy)

ORANGE BERRY IN TRANSVERSE SECTION

FALSE. Part of the receptacle becomes fleshy.

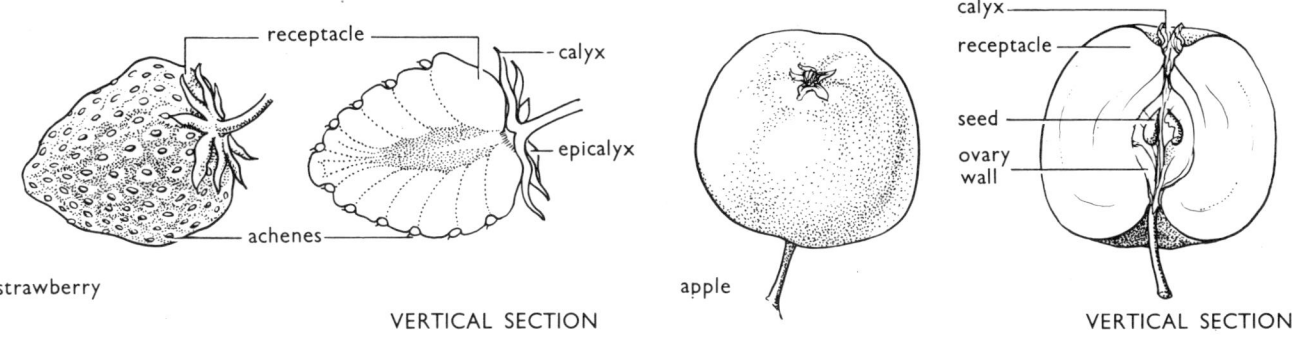

strawberry

VERTICAL SECTION

SWOLLEN RECEPTACLE WITH EXTERNAL TRUE FRUITS

apple

VERTICAL SECTION

SWOLLEN RECEPTACLE ENCLOSING TRUE FRUITS (POME)

Multiple fruits (made from an inflorescence)

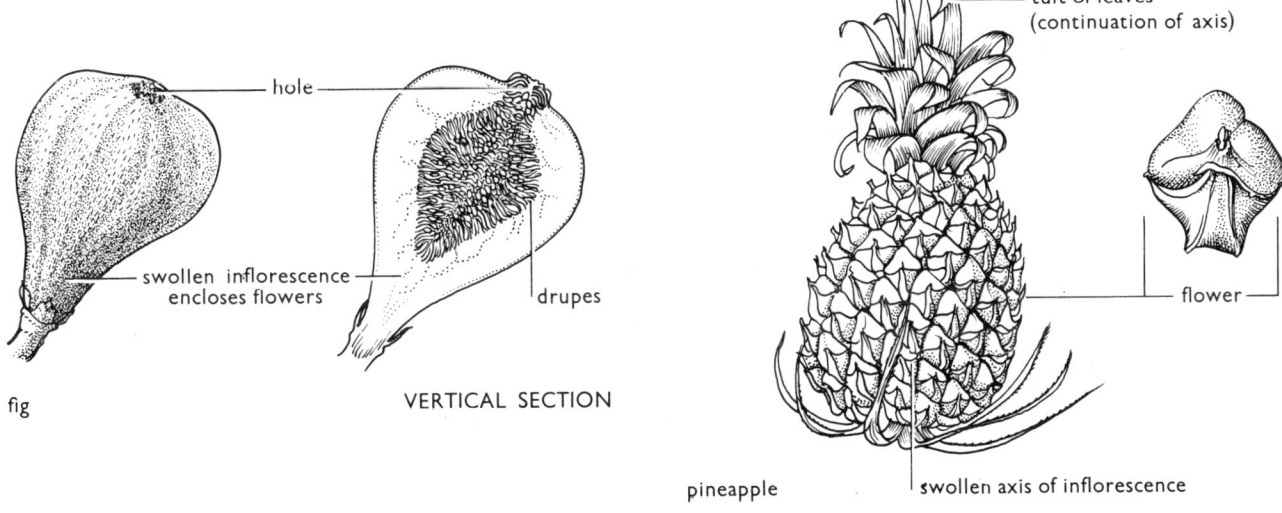

fig

VERTICAL SECTION

pineapple

DISPERSAL OF FRUITS AND SEEDS

DISPERSAL BY ANIMALS

  (i) Hooked fruits which become attached to their bodies. Burdock. Goose grass.
 (ii) Nuts, carried by animals and hoarded in stores. Acorn, Beech.
(iii) Succulent and false fruits. The fruits are eaten but the seeds pass through the digestive systems unharmed, and are egested in the faeces. Plum, Orange, Strawberry, Apple, Fig.

DISPERSAL BY WIND

  (i) Light seeds, e.g. Orchid and Shepherd's Purse. The seeds of Poppy and Campion are dispersed as the capsules move in the wind.
 (ii) Seed parachutes, e.g. Willow Herb.
(iii) Winged seeds, e.g. Pine. (The wing is made from part of the scale of the cone.)
 (iv) Fruit parachutes. The Dandelion.
  (v) Winged fruits, e.g. Ash, Sycamore, Hornbeam, Lime, Elm.

DISPERSAL BY WATER

Water Lily seeds, buoyant due to air spaces inside them.

EXPLOSIVE MECHANISMS

E.g. Gorse, Broom and Pea.

Method of study

Make a collection of as many different seeds and fruits as possible. Study their structure and mode of dispersal.

SEED STRUCTURE

ENDOSPERMIC. Castor Oil (*Ricinus communis*).

WHOLE SEED

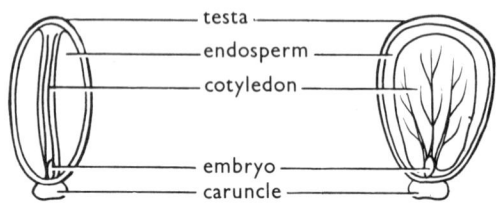

SEED CUT ACROSS NARROWEST SECTION   SEED CUT ACROSS WIDEST SECTION

NON-ENDOSPERMIC. Broad Bean (*Vicia faba*).

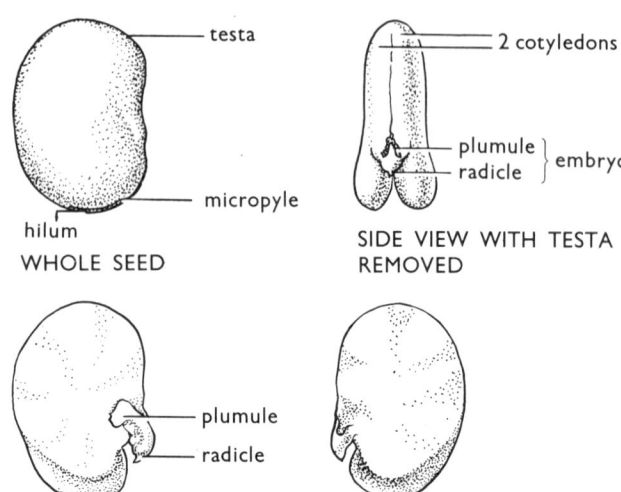

TESTA REMOVED AND COTYLEDONS SEPARATED

GERMINATION OF SEEDS

EPIGEAL. (The cotyledons emerge above ground.) Castor Oil (*Ricinus communis*).

STAGES IN GERMINATION

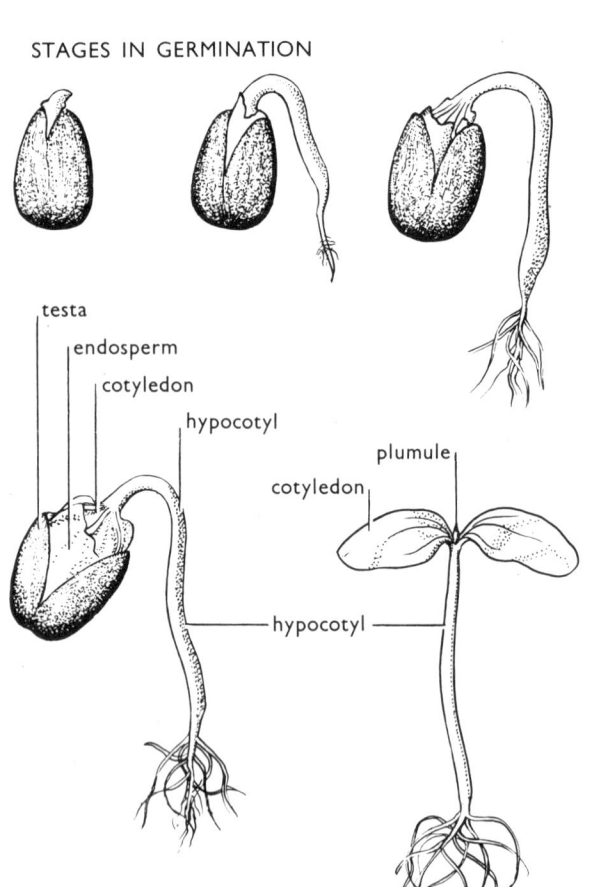

HYPOGEAL. (The cotyledons remain below ground.) Broad Bean (*Vicia faba*).

STAGES IN GERMINATION

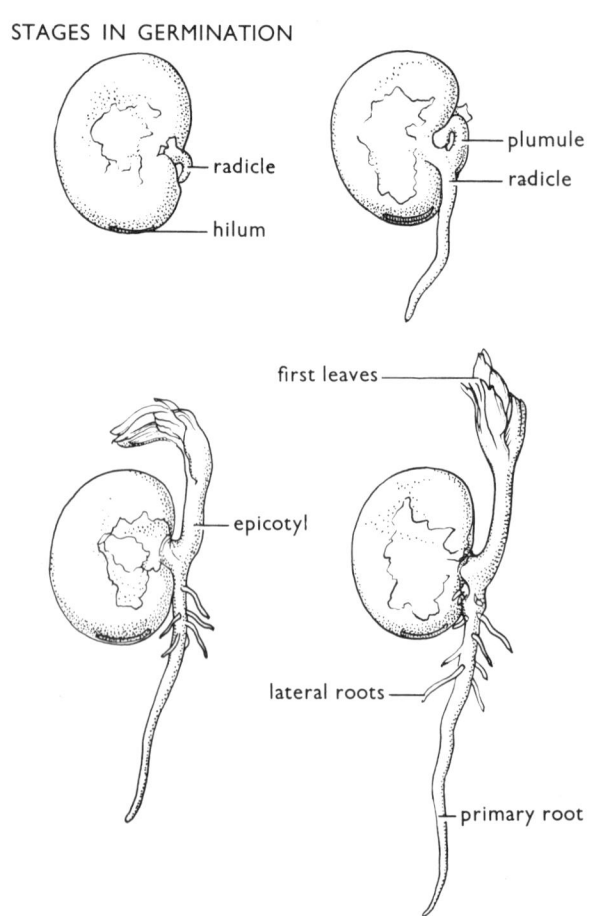

GERMINATION OF FRUITS

EPIGEAL. Sunflower (*Helianthus* spp.).

WHOLE
FRUIT

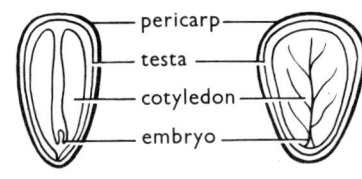

FRUIT CUT
ACROSS
NARROWEST
SECTION

FRUIT CUT
ACROSS
WIDEST
SECTION

pericarp — testa — cotyledon — embryo

STAGES IN GERMINATION

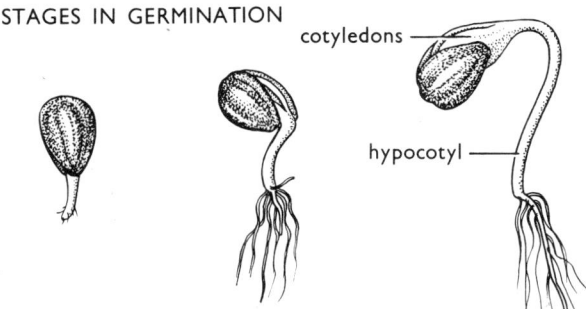

cotyledons

hypocotyl

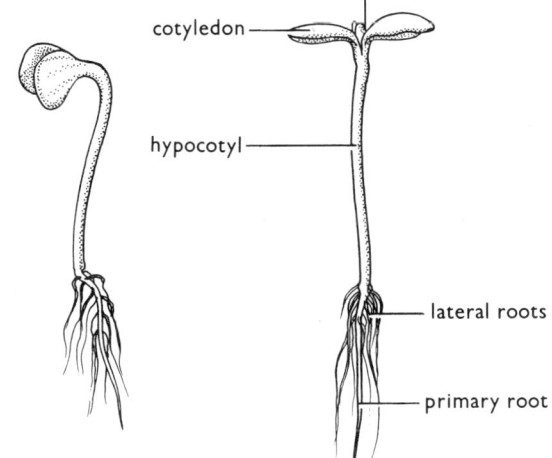

plumule
cotyledon
hypocotyl
lateral roots
primary root

HYPOGEAL. Maize (*Zea mais*).

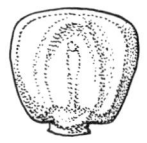

WHOLE FRUIT

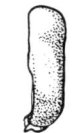

SIDE VIEW

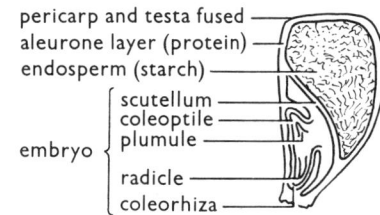

pericarp and testa fused
aleurone layer (protein)
endosperm (starch)
embryo {
  scutellum
  coleoptile
  plumule
  radicle
  coleorhiza
}

FRUIT CUT ACROSS
NARROWEST SECTION

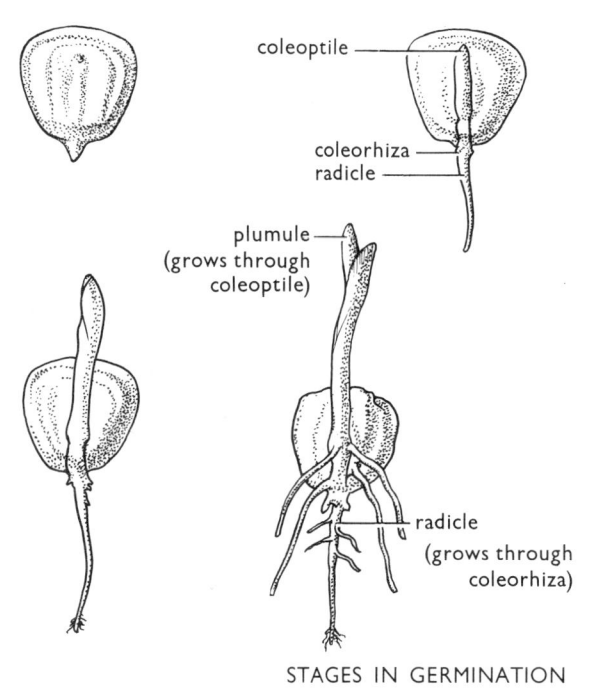

coleoptile
coleorhiza
radicle
plumule
(grows through
coleoptile)
radicle
(grows through
coleorhiza)

STAGES IN GERMINATION

GERMINATION

Types of seeds

Endospermic. The food store is in the *endosperm*, a tissue which develops in the embryo sac by the division of the central nucleus after the second male gamete in the pollen tube has fused with it.

Non-endospermic. The food store is contained in the cotyledons.

Types of germination

Epigeal. The cotyledons push above the ground at an early stage in the growth of the seed, and they become photosynthetic.

Hypogeal. The cotyledons remain below ground. The plumule pushes above ground and is the first structure to become photosynthetic.

Conditions necessary for germination

After a seed has completed its period of dormancy, it will germinate if the following conditions are satisfied.

(i) There is an adequate water supply to mobilize the food reserves inside the seed. Water enters through the *micropyle* and *testa*.

(ii) There is an adequate supply of oxygen for the respiratory requirements of the seed.

(iii) There is a suitable temperature, the optimum for most seeds being approximately 25°C.

Method of examination

Germinate a variety of seeds and examine the stages in their development at two-day intervals.

17.2.4 GROWTH

Development of the Horse Chestnut (*Aesculus* spp.)

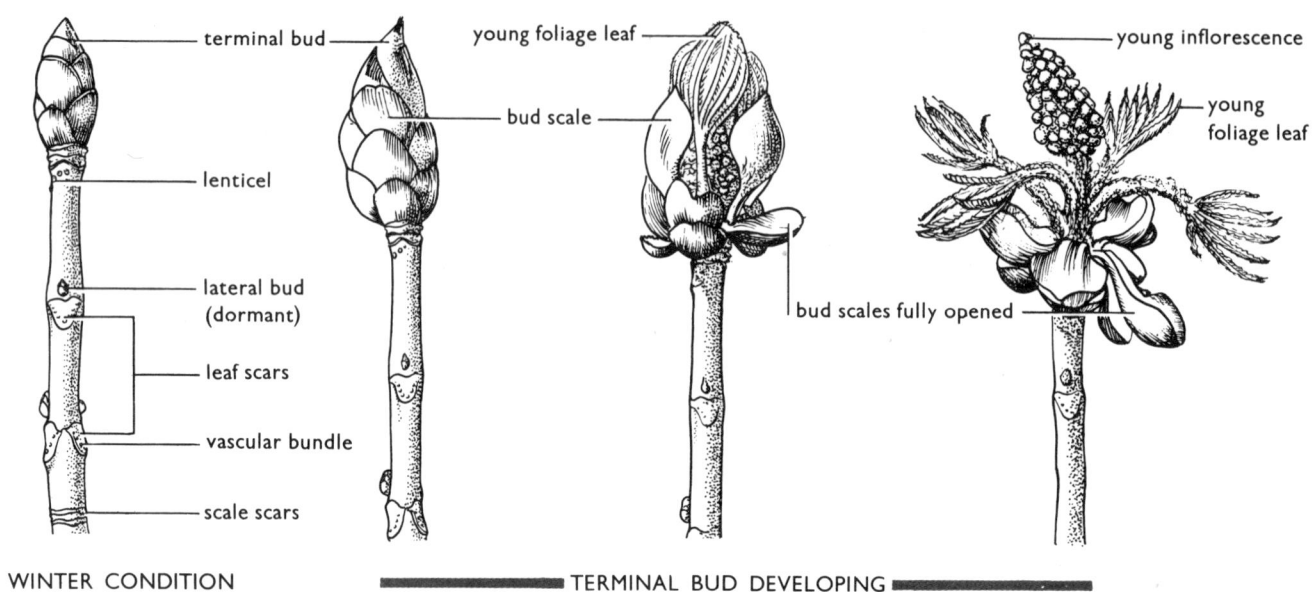

WINTER CONDITION ◼◼◼◼◼◼ TERMINAL BUD DEVELOPING ◼◼◼◼◼◼

Winter Stages of Common Twigs

ELM   OAK   BEECH   COMMON ASH   SYCAMORE

Leaves

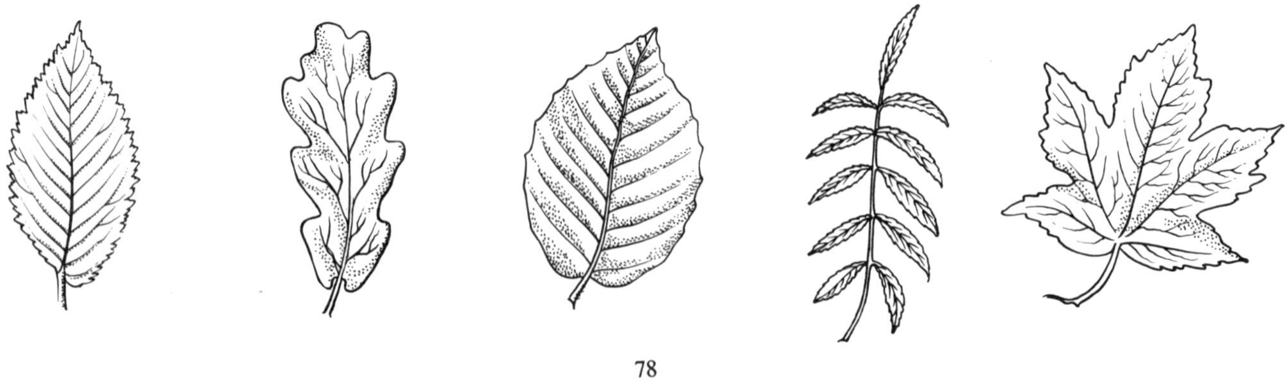

17.2.5 VEGETATIVE REPRODUCTION

(a) Natural

CREEPER (horizontal shoot above ground)

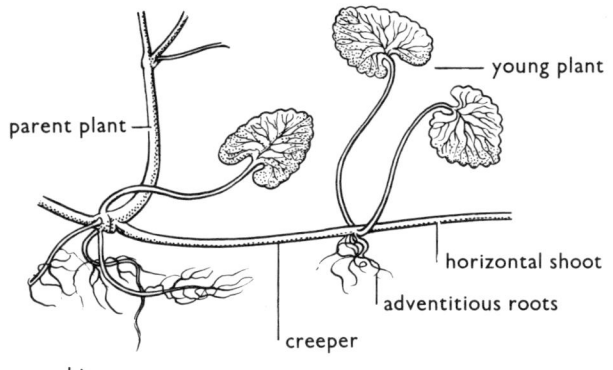

ground ivy

RUNNER (horizontal shoot above ground)

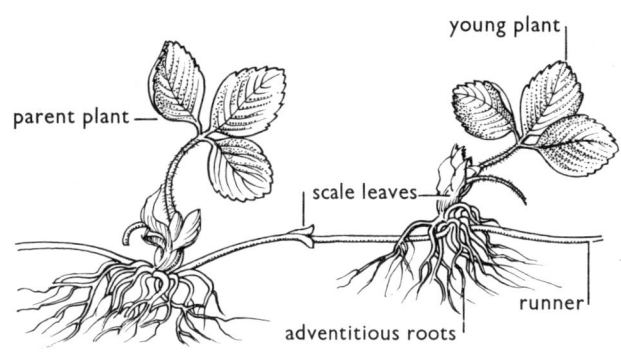

strawberry

SUCKER (horizontal shoot below ground)

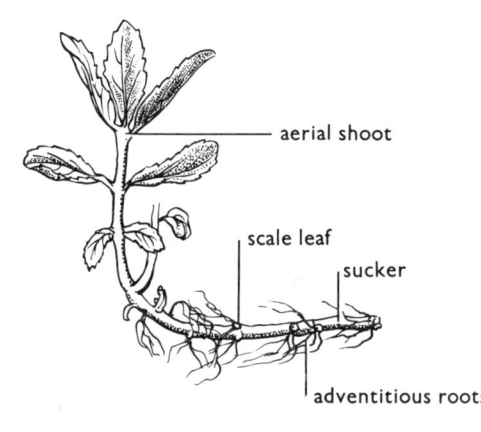

mint

STOLON (swollen terminal bud)

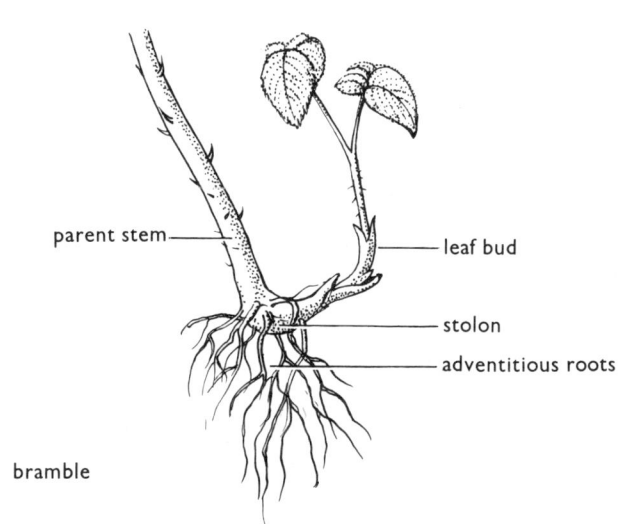

bramble

(b) Artificial

CUTTINGS (lateral shoot cut below a node)

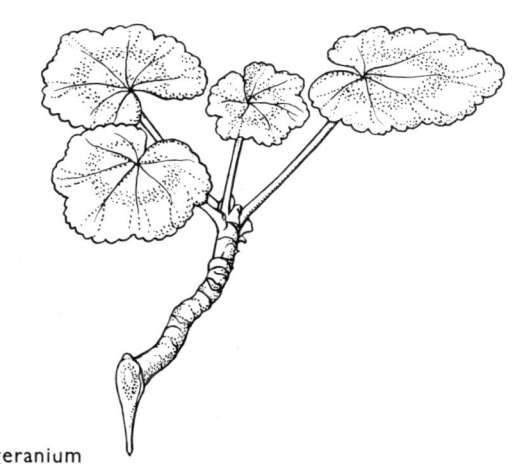

geranium

GRAFTINGS

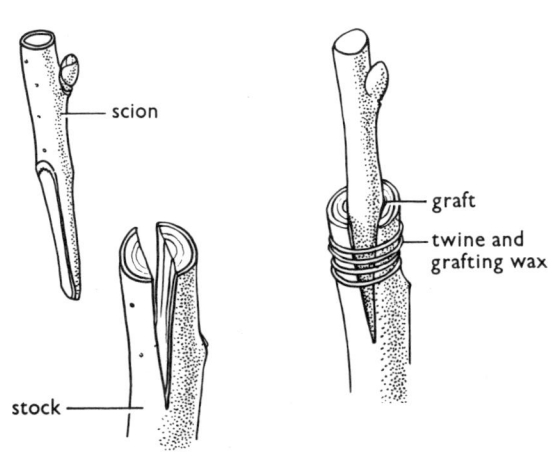

17.2.6 FOOD STORAGE

Stems

(a) Swollen stem tips

TUBERS

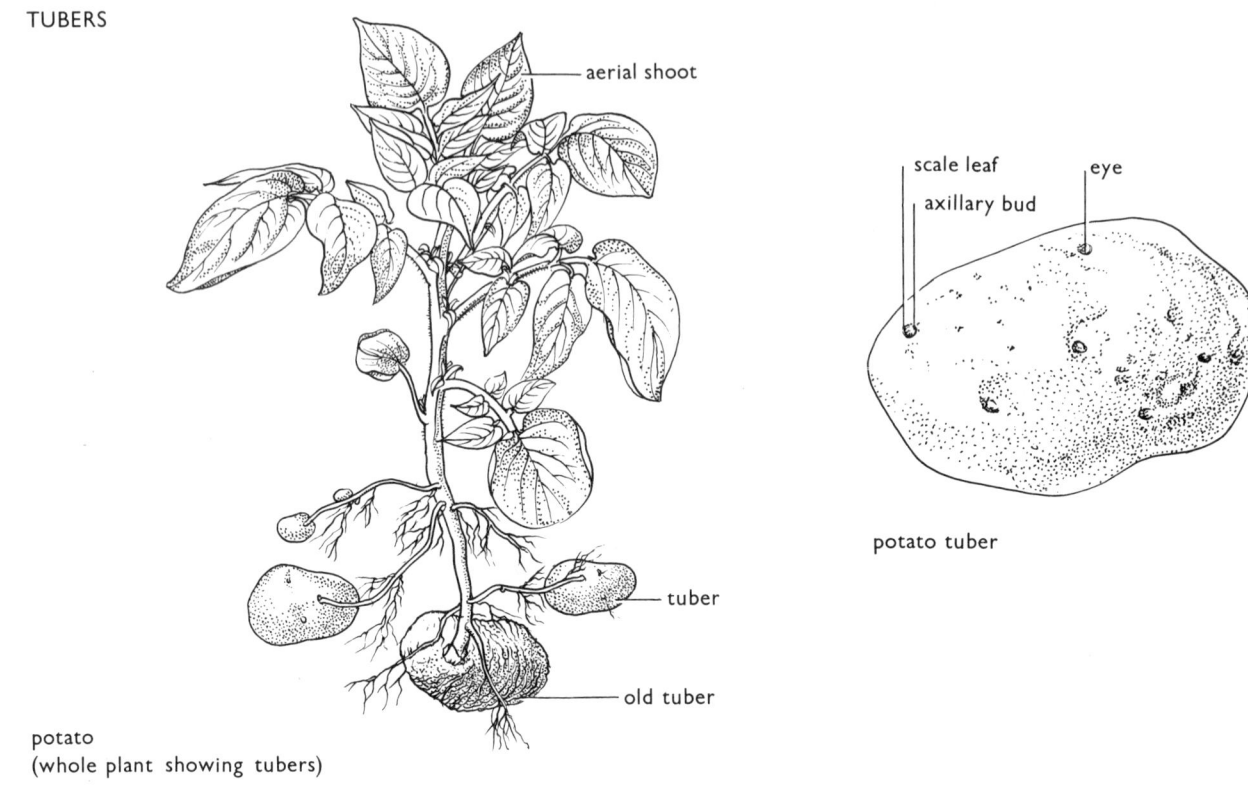

potato
(whole plant showing tubers)

potato tuber

(b) Swollen stem bases

CORMS

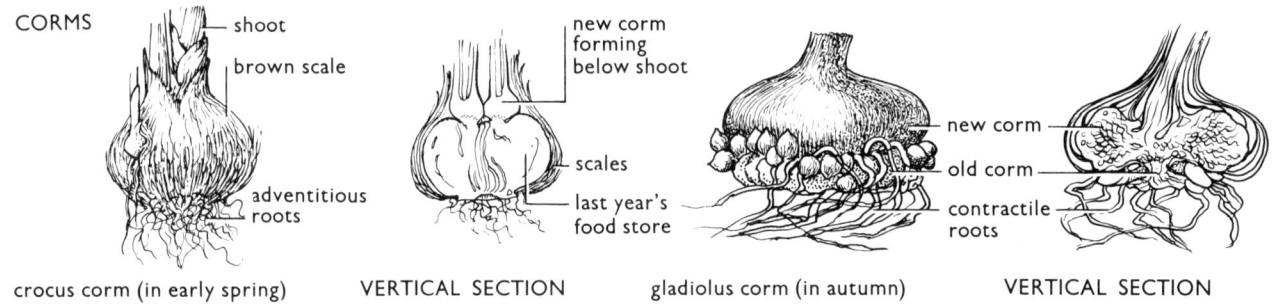

crocus corm (in early spring)   VERTICAL SECTION   gladiolus corm (in autumn)   VERTICAL SECTION

(c) Horizontal stems

RHIZOMES

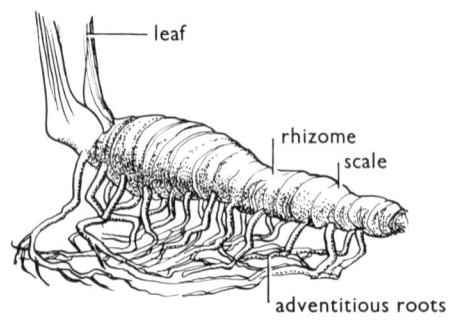

iris

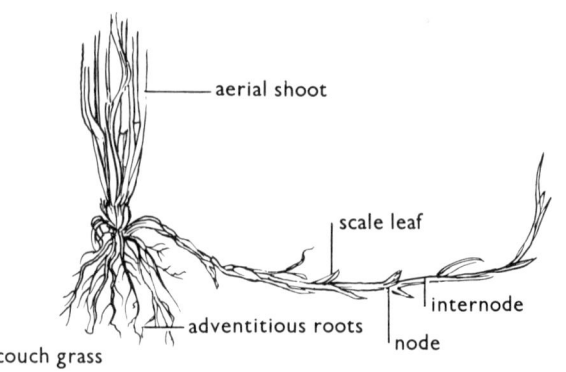

couch grass

(*d*) **Swollen stem axis**

brussels sprout

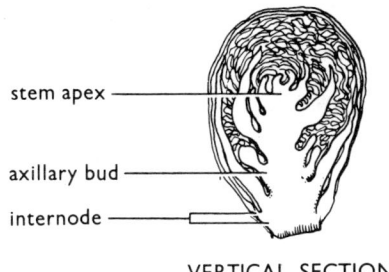

stem apex

axillary bud

internode

VERTICAL SECTION

Leaves (swollen leaf bases)

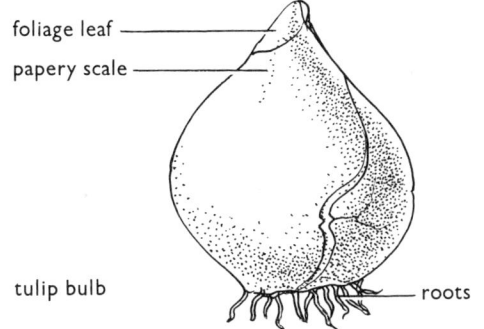

foliage leaf

papery scale

tulip bulb

roots

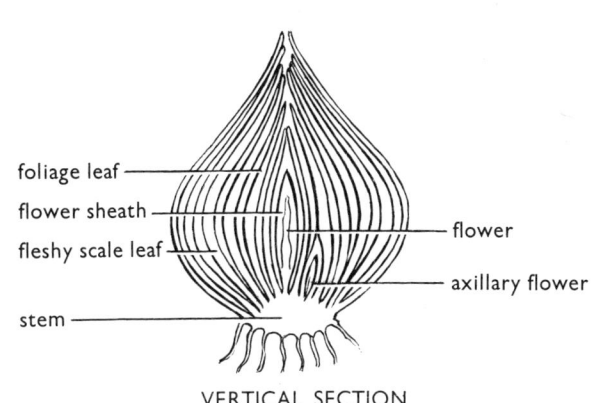

foliage leaf

flower sheath

fleshy scale leaf

flower

axillary flower

stem

VERTICAL SECTION

Roots
(*a*) Tap roots

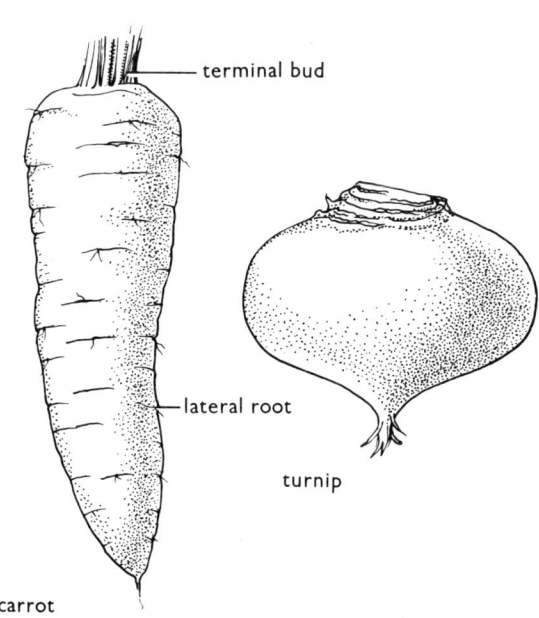

terminal bud

lateral root

carrot

turnip

(*b*) Tuberous roots

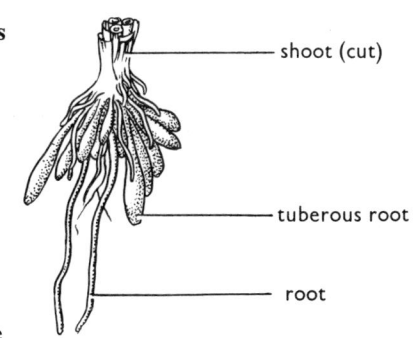

shoot (cut)

tuberous root

root

lesser celandine

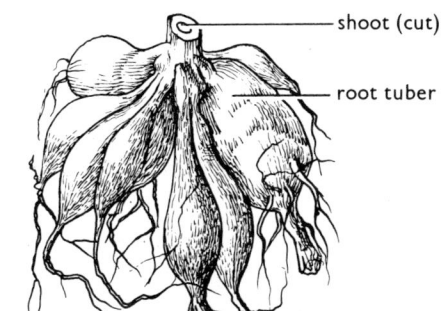

shoot (cut)

root tuber

dahlia

17.2.7 ABNORMAL METHODS OF FEEDING

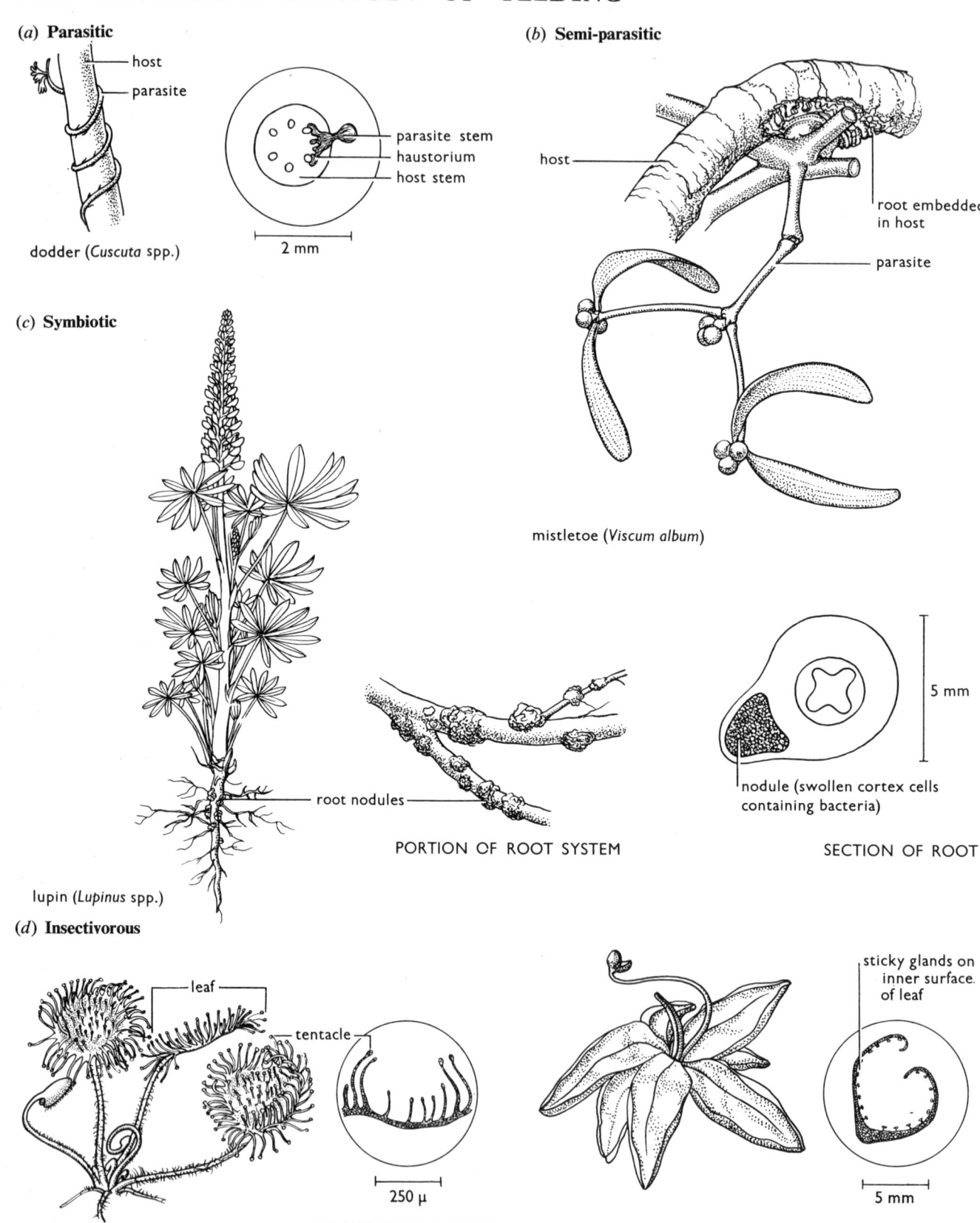

(a) **Parasitic**

host
parasite

parasite stem
haustorium
host stem

2 mm

dodder (*Cuscuta* spp.)

(b) **Semi-parasitic**

host

root embedded in host

parasite

mistletoe (*Viscum album*)

(c) **Symbiotic**

root nodules

PORTION OF ROOT SYSTEM

5 mm

nodule (swollen cortex cells containing bacteria)

SECTION OF ROOT

lupin (*Lupinus* spp.)

(d) **Insectivorous**

leaf
tentacle

250 μ

TRANSVERSE SECTION OF LEAF

sundew (*Drosera rotundifolia*)

sticky glands on inner surface of leaf

5 mm

TRANSVERSE SECTION OF LEAF

butterwort (*Pinguicula* spp.)

17.2.8 ADAPTATION TO ENVIRONMENT

(a) **Hydrophytes** Living in water.

FLOATING LEAVES

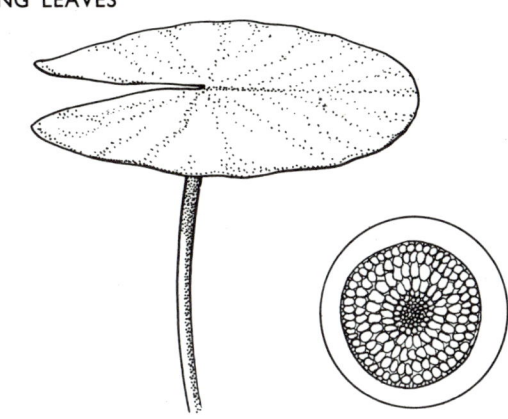

water lily (*Nymphaea* spp.)

section of stem to show air spaces

FLOATING PLANTS

duckweed (*Lemna* spp.)

frond

root

(b) **Xerophytes** Modifications for cutting down the rate of water loss.

WATER STORING LEAVES

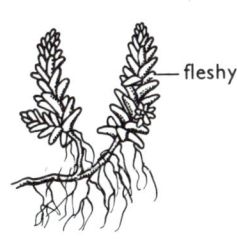

fleshy leaf

stonecrop (*Sedum* spp.)

THICK WAXY CUTICLE

holly (*Ilex* spp.)

HAIRY LEAF

mullein (*Verbascum* spp.)

NARROW LEAVES WITH SUNKEN STOMATA

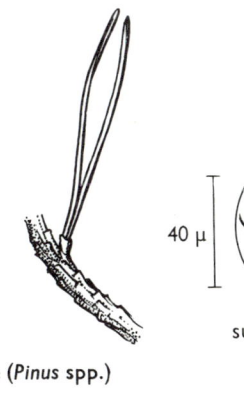

pine (*Pinus* spp.)

stoma

epidermis cell

guard cell

40 μ

sunken stoma under high power

ROLLED LEAVES

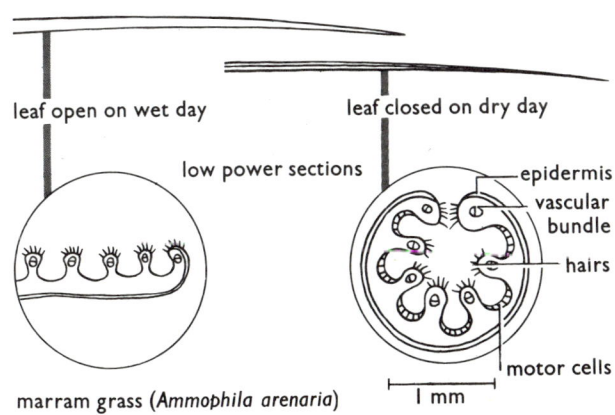

leaf open on wet day

leaf closed on dry day

low power sections

epidermis

vascular bundle

hairs

motor cells

marram grass (*Ammophila arenaria*)

1 mm

REDUCED LEAF SURFACE

heather (*Erica* spp.)

gorse (*Ulex* spp.)

(*c*) **Halophytes**   Living in saline conditions.

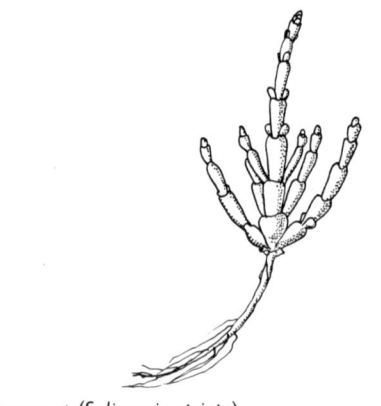

glasswort (*Salicornia stricta*)

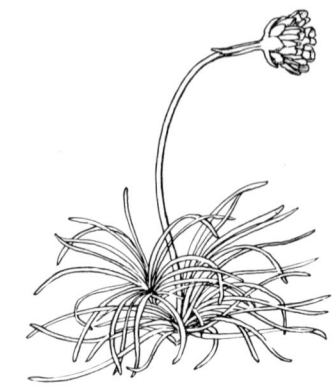

thrift (*Armeria maritima*)

(*d*) **Climbers**

CURVED THORNS

blackberry (*Rubus* spp.)

TWINING

bindweed (*Convolvulus* spp.)

CURVED HOOKS

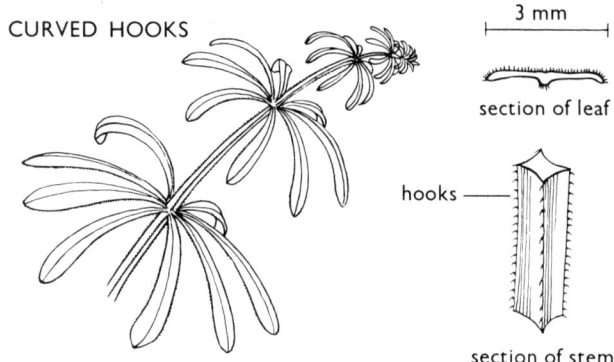

3 mm

section of leaf

hooks

section of stem

goose grass (*Galium aparine*)

TENDRILS

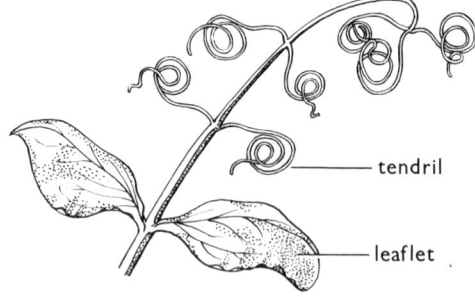

tendril

leaflet

pea (*Lathyrus* spp.)

ROOT CLIMBER

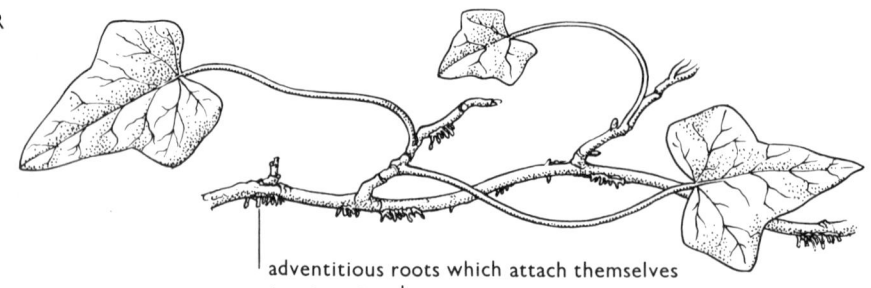

adventitious roots which attach themselves
to a tree trunk

ivy (*Hedera helix*)

17.2.9 MICROSCOPIC STRUCTURE

CELLS

POTATO CELLS

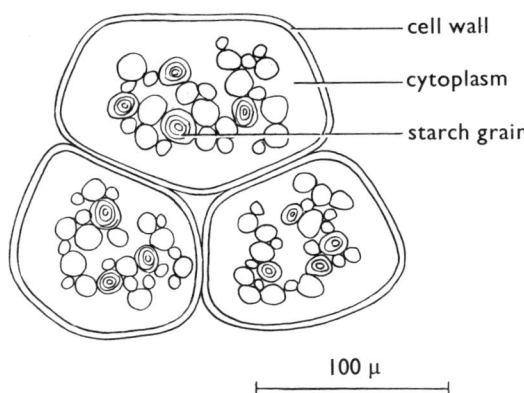

cell wall

cytoplasm

starch grain

100 μ

ONION CELLS

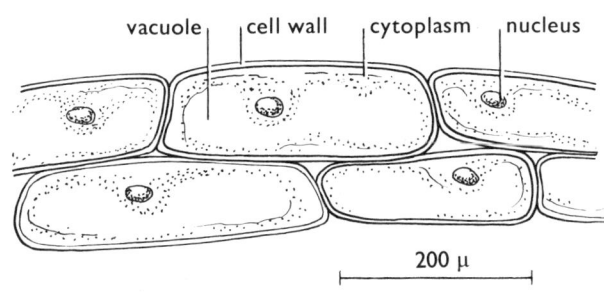

vacuole | cell wall | cytoplasm | nucleus

200 μ

Potato cells

The potato is a swollen stem tip and a large proportion of its tissue consists of comparatively large cells. They are irregular in shape and are surrounded by a cellulose cell wall. Inside the cells are prominent starch grains which constitute the food store.

Obtain a small cylinder of tissue with a cork borer and then cut a few thin transverse sections using a razor blade. Mount the thinnest section in a drop of water on a slide and place a coverslip over it. Place a drop of iodine at one edge of the coverslip and draw the fluid underneath with a piece of filter paper positioned at the side opposite to the iodine spot. Observe the colour change in the starch grains.

Onion cells

Remove one of the inner scales from a fresh onion, and using a pair of forceps tear off a small piece of epidermis from its surface. Mount a small piece in a drop of iodine on a slide, cover it with a coverslip and examine it under the high power of the microscope. The cells are more regular in outline than the cells of potato tissue, the iodine stains cellulose and nuclear material yellow.

TRADESCANTIA VIRGINICA

Tradescantia stamen hair cells

Obtain some fresh flowers of *Tradescantia virginica*. Remove a stamen and place it on a white background and identify the purple hairs covering the sides of the filament. Pick a few of these hairs off with a pair of forceps and mount them immediately on a slide in a drop of water. Place a coverslip over them and focus a cell under the high power of the microscope keeping the illumination low by adjusting the iris diaphragm. Observe the streaming movements along well-defined channels in the cytoplasm.

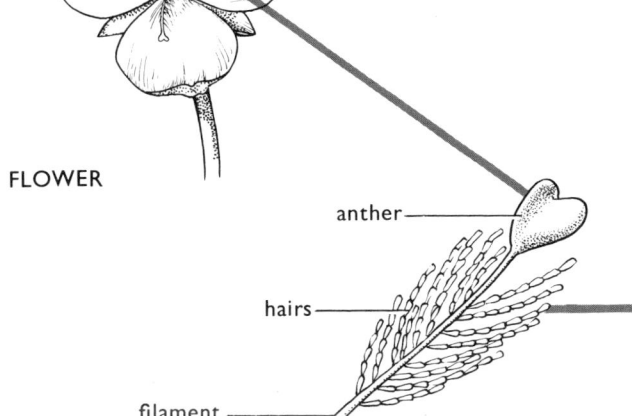

FLOWER

anther

hairs

filament

STAMEN

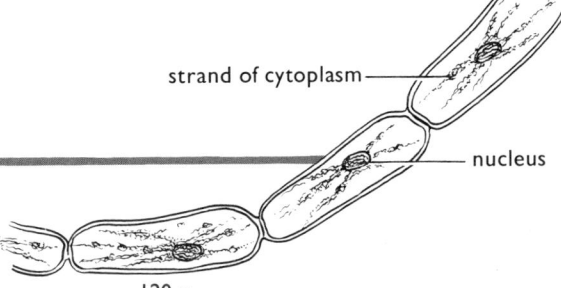

cell wall

strand of cytoplasm

nucleus

120 μ

CELLS FROM STAMEN HAIR

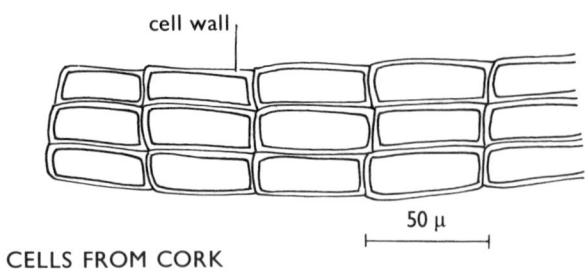

cell wall

50 μ

CELLS FROM CORK

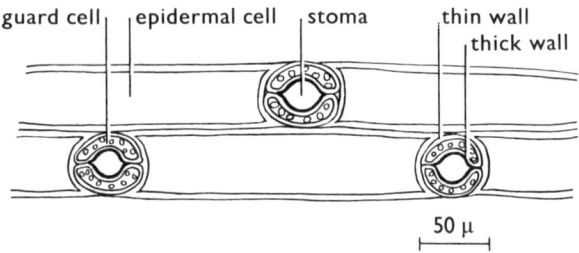

guard cell | epidermal cell | stoma | thin wall | thick wall

50 μ

CELLS FROM EPIDERMIS OF IRIS

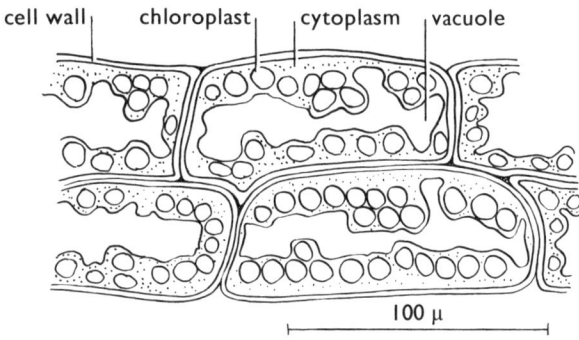

cell wall | chloroplast | cytoplasm | vacuole

100 μ

CELLS FROM LEAF OF ELODEA

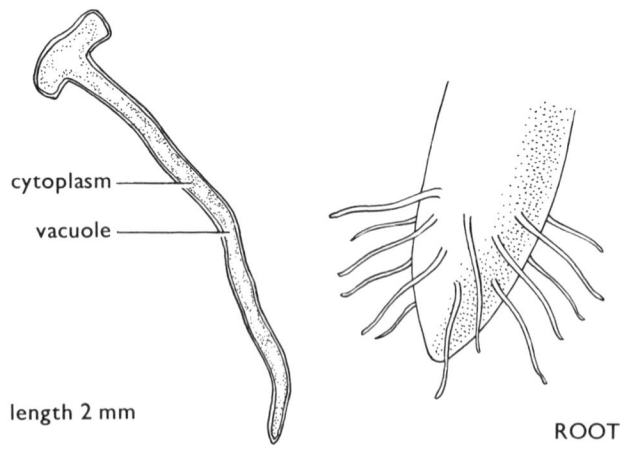

cytoplasm

vacuole

length 2 mm

ROOT

ROOT HAIR CELL FROM CRESS

Cork cells

The tissue in the bark of trees consists mainly of rectangular cells made by a special cork cambium in the outer part of the cortex. The walls of these cells are strengthened to make them water-proof and gas-proof with a material called *suberin*, and cork tissue as a whole protects the outside of woody stems and roots. Cork cells were probably the first cells to be observed by Robert Hooke (1635–1703).

Obtain a small piece of bark from lime or elderberry, cut thin sections and mount them in a drop of iodine on a slide. Cover them with a coverslip and examine them under the high power of the microscope.

Iris leaf cells

Peel off a thin layer of the epidermis from the leaf of an Iris plant, and mount a small piece in a drop of water on a microscope slide. Examine it under the low and high power and note the regularly arranged elongated cells with stomata systematically dispersed amongst them. This tissue can also be used for studying the method by which stomata open and close. Observe the effect of replacing the water with a weak sucrose solution, and subsequently with a strong sucrose solution.

Canadian Pondweed leaf cells

This water plant (*Elodea canadensis*) has small leaves about 1 cm long attached closely to its submerged stem. Using a pair of forceps remove a complete leaf and mount it in a drop of water on a slide and cover it with a coverslip. As the leaf is comparatively thin, cells are easily visible under the microscope; focus one under the high power and observe the distribution of the chloroplasts in the cell cytoplasm. Frequently these can be seen moving inside the living cell, especially when brightly illuminated.

Cress root hair cells

Grow some Cress seeds until the seedlings develop to a length of a few cm. Use one seedling to examine the root hair cells which are formed in the growth region just behind the tip of the seedling root. Cut off the terminal cm of the root and mount it in a drop of water on a slide. Cover with a coverslip and examine some typical root hair cells under the low and high power of the microscope. Suck the water from one edge of the coverslip with a piece of dry filter paper, and replace it with a drop of iodine introduced from a pipette. This will make it easier to observe the detailed structure of the cell.

16.2   The Marsh Horsetail (*Equisetum palustre*)

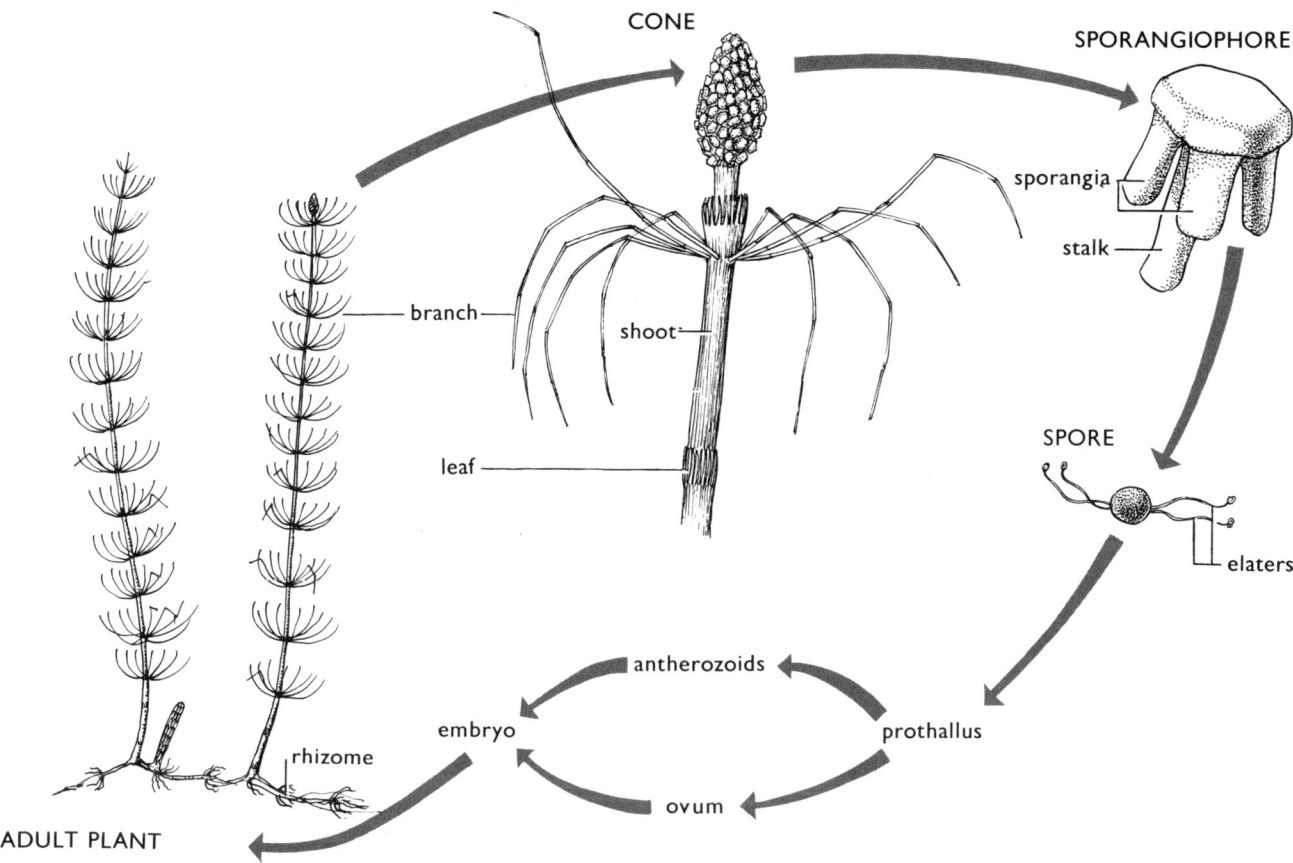

Habitat

In damp marshy places.

Structure

There is a vertical ribbed stem which is the main photosynthetic organ, as the leaves are minute spikes joined at their bases and occurring in whorls at each of the prominent nodes. Branches also arise at many of the nodes. The clear distinction between nodes and internodes gives a jointed appearance, which is responsible for the name Arthrophyta. The stems arise from horizontal rhizomes underground, which bear whorls of threadlike roots.

Reproduction

In Spring, cone-like structures appear at the tips of certain shoots. They consist of numerous *sporangiophores* which are somewhat like a mushroom in shape. Each *sporangiophore* carries 5–10 *sporangia* directed towards the main axis of the cone. In these, green spores develop, which when ripe are helped to disperse by 'flicking cells' (*elaters*) which also develop in the *sporangium*, and whose shape varies rapidly with changes of humidity. The spores germinate when they fall on damp soil, and grow into gametophytes in the form of small green *prothalli*, which are attached to the soil by *rhizoids* from the lower surface. Reproductive organs develop in due course; the males are simple *antheridia* inside which *antherozoids* are produced. The females are flask-shaped *archegonia*. Fertilization takes place when *antherozoids* emerge from the *antheridium* and swim to the ovum inside each *archegonium*. The fertilized zygote develops into an embryo which starts its life as a parasite on the gametophyte, but when it grows above the soil level it becomes photosynthetic, and develops into an independent sporophyte.

Method of examination

1. Examine a fully grown sporophyte and record the structure of the shoot, rhizome and roots.
2. Examine the structure of a ripe cone. Remove one of the sporangiophores and study its structure with the aid of a hand lens.
3. Remove a sporangium from the sporangiophore. Squash it on a dry microscope slide, and examine it under the low power. If elaters are present, breathe gently on to the slide, and observe their movement as the humidity changes.

14.7 PLANT PLANKTON

The word plankton is aptly taken from the Greek, and describes organisms which float in water at the mercy of the currents. The majority are less than 1 mm in size, and the plant plankton (*Phytoplankton*) consists of Algae whose cells often have projections acting as buoyancy devices.

Diatoms have a great variety of species characterized by the possession of siliceous cell walls in two halves which overlap each other, and there are often spiny extensions acting as devices to increase buoyancy. Inside the protoplasm orange coloured chloroplasts occur. Although they are plants, many are able to perform gliding movements which are quite spectacular in some cases (e.g. *Bacillaria*). They are common in fresh and sea water.

Desmids have cellulose cell walls which are often double in appearance like those of *Diatoms*. They are common in acid peaty waters.

Dinoflagellates have cellulose cell walls and move by means of flagellae. They occur in fresh and sea water.

The organisms of *Phytoplankton* are responsible for most of the photosynthetic activity in the surface layers of ponds, lakes and the sea. In the Northern Hemisphere the *Phytoplankton* develops in early Spring as the day length increases. Apart from light, there are important chemicals which they require for growth, especially nitrates and phosphates. These are normally brought from the depths where they have accumulated over Winter, to the surface by means of currents. As the surface waters get warmer in early Summer the metabolic activities of the plant population increase rapidly to utilize much of the nitrate and phosphate. In consequence stocks of these chemicals are depleted. At the same time animal larvae develop using the *Phytoplankton* as the main source of their food so that as Summer progresses, the plant population declines. As the chemical supply fails, most of the animals together with dead plant remains drop to lower levels. Sometimes another short cycle occurs in early Autumn if fresh supplies of chemicals are brought to the surface so causing a new flush of *Phytoplankton*; otherwise there is little activity until chemicals are brought to the surface again in the following Spring. So, the ultimate source of food for aquatic animals is the *Phytoplankton* which in turn is dependent on sunlight energy for the processes of photosynthesis (See page 105).

Some authorities quote that the *Phytoplankton* in clear tropical seas can fix 1 kg of carbohydrate per square meter per year, which would give about 10 tons of carbohydrate per acre per year. Such calculations are difficult to make with accuracy and in the English Channel a figure of 17 cwt per acre per year is said to be near the truth.

The study of plankton can only be carried out satisfactorily on fresh material, as a great deal of its character is lost if attempts are made to preserve it. Organisms can be caught with a fine mesh plankton net towed slowly through the water and should be stored in a thermos flask until they are ready for examination. Good fresh water specimens can often be obtained more easily on the sides of tanks and small pools. They can be sucked off this habitat with a small pipette. Examine fresh material under the low and high powers of the microscope.

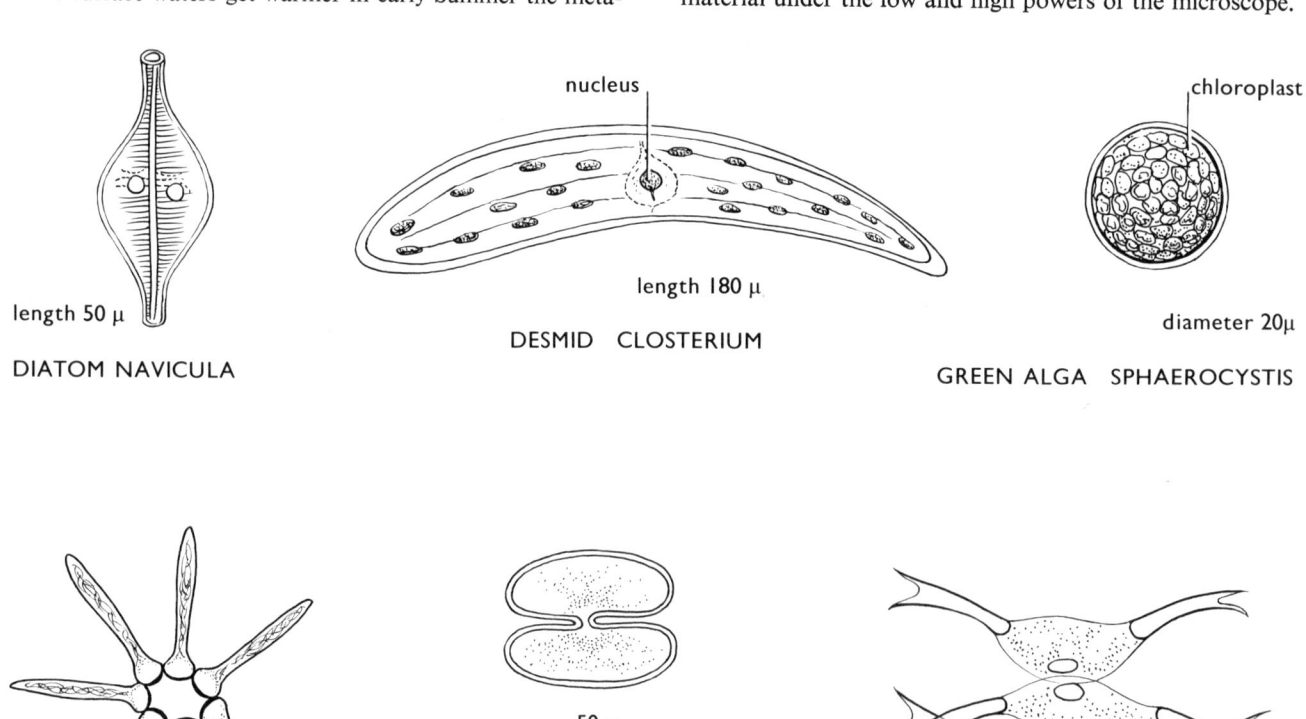

length 50 μ

DIATOM NAVICULA

nucleus

length 180 μ

DESMID CLOSTERIUM

chloroplast

diameter 20μ

GREEN ALGA SPHAEROCYSTIS

length 50 μ

DIATOM ASTERIONELLA

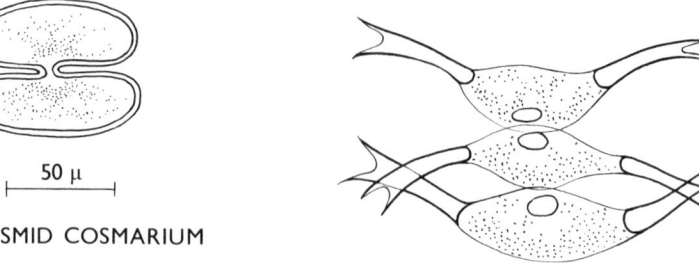

50 μ

DESMID COSMARIUM

length 50 μ

DESMID STAURASTRUM

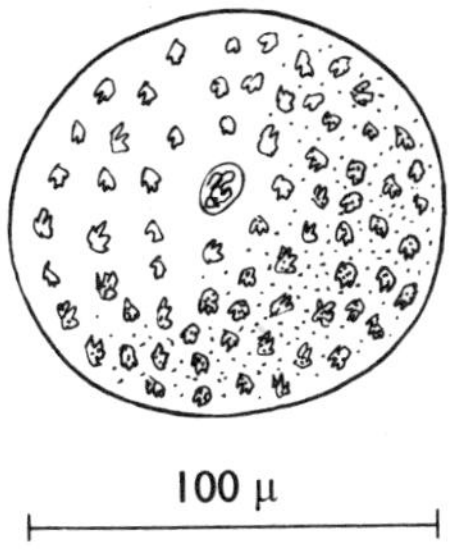

DIATOM HALOSPHAERIA

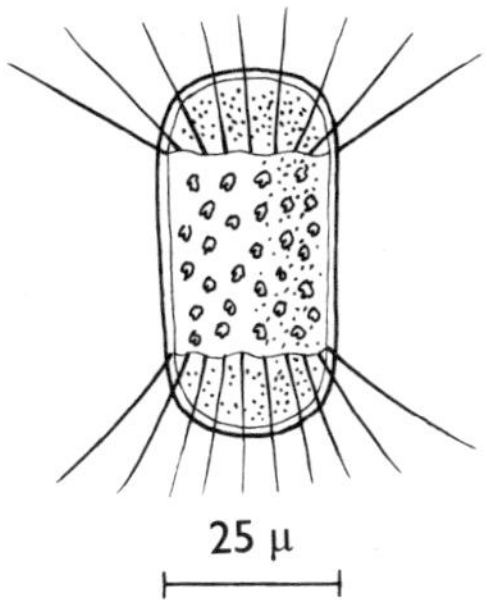

100 μ

DIATOM CORETHRON

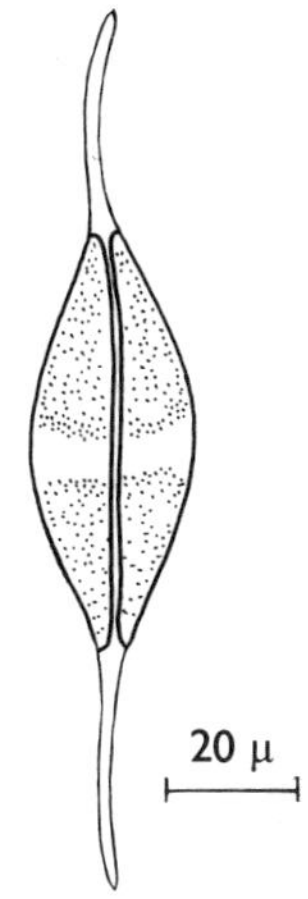

25 μ

DIATOM NITZSCHIA

20 μ

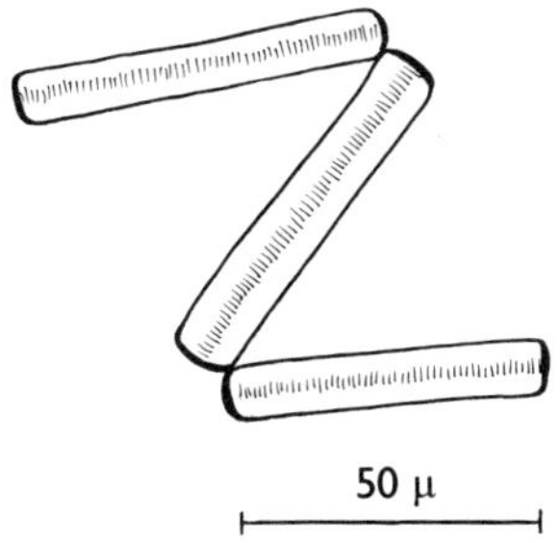

DIATOM THALASSIOTHRIX

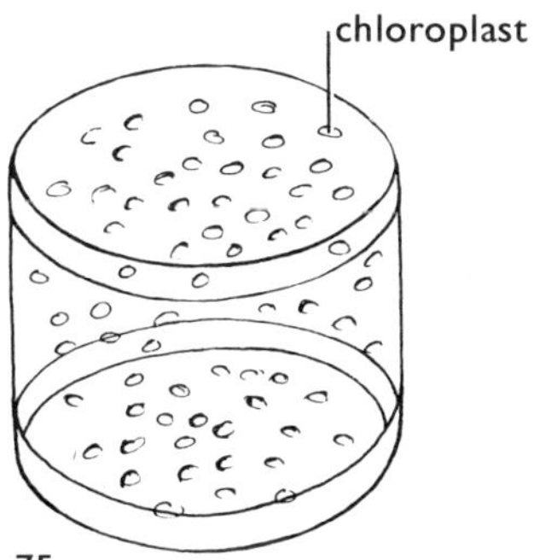

50 μ

chloroplast

diameter 75 μ

DIATOM COSCINODISCUS

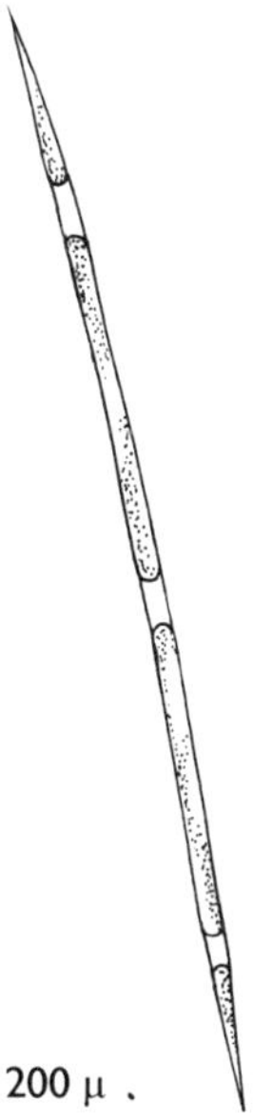

length 200 μ .

DIATOM RHIZOSOLENIA

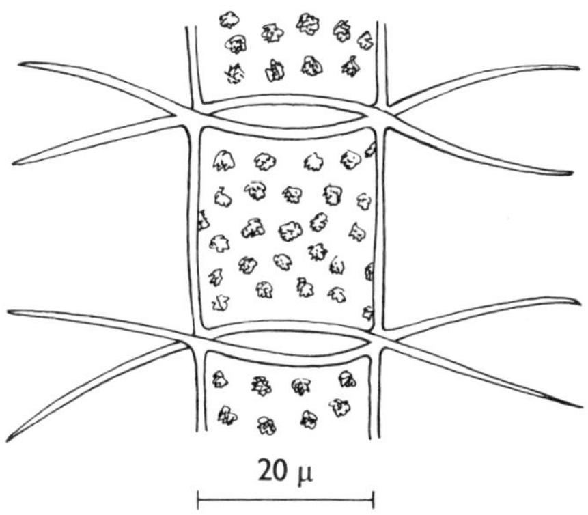

DIATOM CHAETOCEROS

20 μ

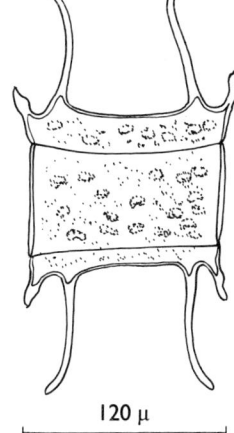

DIATOM BIDDULPHIA

120 μ

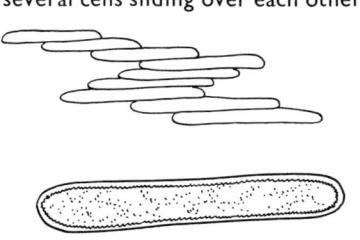

several cells sliding over each other

length 50 μ

DIATOM BACILLARIA

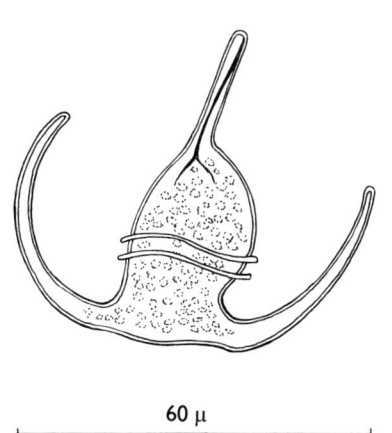

DINOFLAGELLATE CERATIUM

60 μ

Further Animal Types

ANNELIDA

CLASS POLYCHAETA

9.2 The Marine Ragworm (*Nereis diversicolor*)

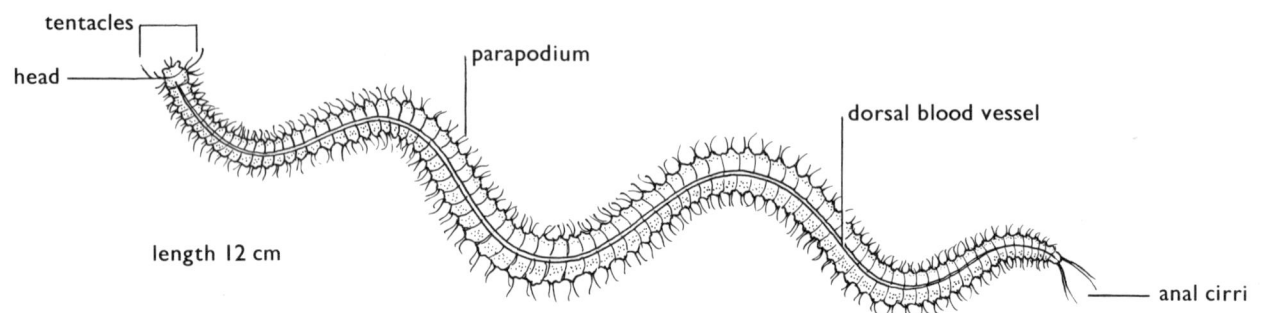

This belongs to the group of Annelid worms known as Polychaeta. Like the earthworm it is segmented and each segment bears a pair of paddle-like appendages called *parapodia* from which numerous bristles project. There is a well-developed head with prominent appendages, and the worm swims with a characteristic side-to-side rippling movement. When feeding it everts its proboscis, revealing a pair of mandibles which seize hold of the prey. There is a prominent dorsal blood vessel. In the process of reproduction sperms and eggs are shed into the sea where fertilization takes place. The resulting zygote develops into a *trochophore* larva, which floats in the plankton. It grows into a *nectochaeta* larva (See page 97) which ultimately sinks in the sea and develops into an adult worm.

9.3 The Lugworm (*Arenicola marina*)

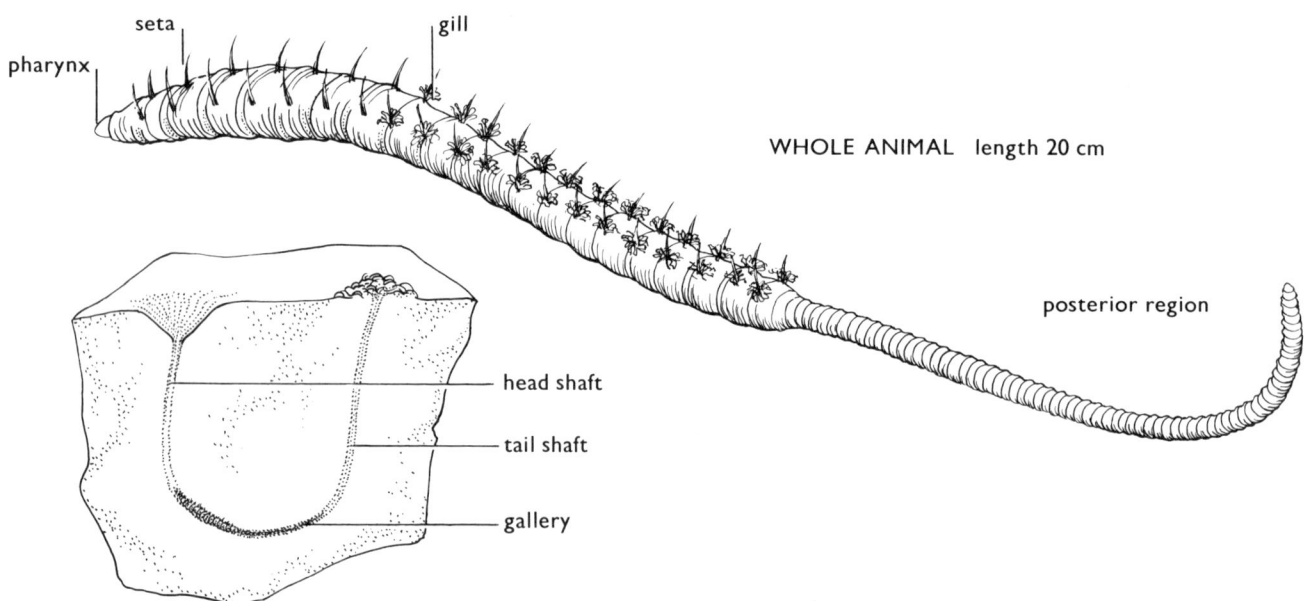

WHOLE ANIMAL   length 20 cm

DIAGRAMMATIC SECTION OF BURROW

At low tide, there are many thousands of casts and depressions, mainly on the lower third of the shore, caused by the animal, which is a deposit feeder and extracts food from the large quantity of sand which passes through its gut. Most of the time it stays at the bottom of its U-shaped burrow, about one foot below the surface. The depression and cast on the sand mark the position of the head and tail shafts of the burrow. When covered with water the worm sets up currents that cause sand from near the surface to fall into the head shaft. It swallows this sand, and at regular intervals of about forty minutes moves backwards up the tail shaft to shoot out another cylinder on to its heap of castings. There are 13 pairs of bright red gills towards the anterior end of the body.

Place a worm in a glass U-tube containing sand, and immerse it in a small tank of sea water. Observe its movements over a period of time.

9.4 Marine Tubeworm (*Pomatoceros triqueter*)

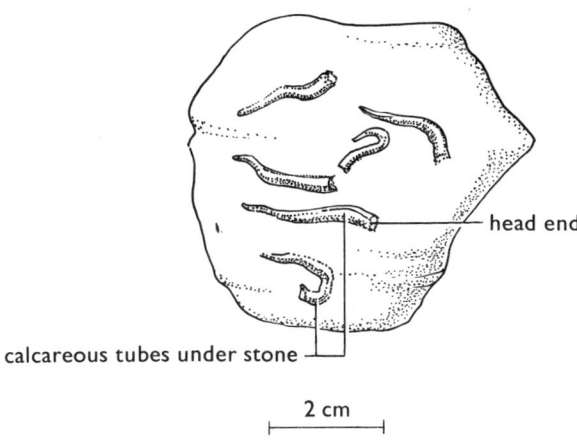

calcareous tubes under stone

head end

2 cm

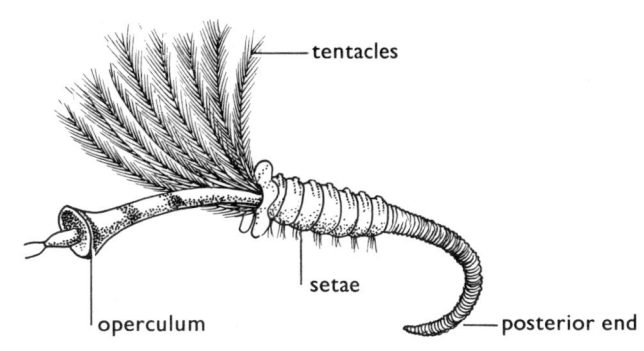

tentacles

operculum

setae

posterior end

MALE REMOVED FROM TUBE

This worm lives in a calcareous tube, which it secretes itself. They are found in large numbers attached to stones on the lower zones of the sea shore. In section each tube is triangular, sharp at one end and blunt at the other, and may reach from 2 to 12 cm in length. The worm is a filter feeder, drawing food into the tube by ciliated tentacles at the anterior end of its body. One tentacle is modified to form an operculum, which stops up the mouth of the tube when the animal withdraws

inside it. The sexes are separate, and gametes are shed into the sea where fertilization takes place.

Place a small stone with *Pomatoceros* tubes in a dish of sea water and examine the worms with a lens when they emerge. Force a worm out of its tube with a seeker and examine in it a clock glass of sea water under the low power of the microscope. These animals can be used for experiments on fertilization, see Section 23.3.3 (iv) for details.

CLASS OLIGOCHAETA

9.6 Other Earthworms

Apart from the Common Worm (*Lumbricus terrestris*) (See page 18) other species are frequently found in soil and compost heaps. The following are typical examples.

CLASS HIRUDINEA

9.5 The Horse Leech (*Haemopsis sanguisuga*)

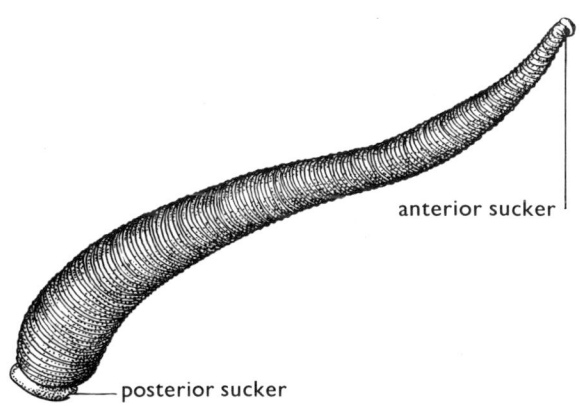

anterior sucker

posterior sucker

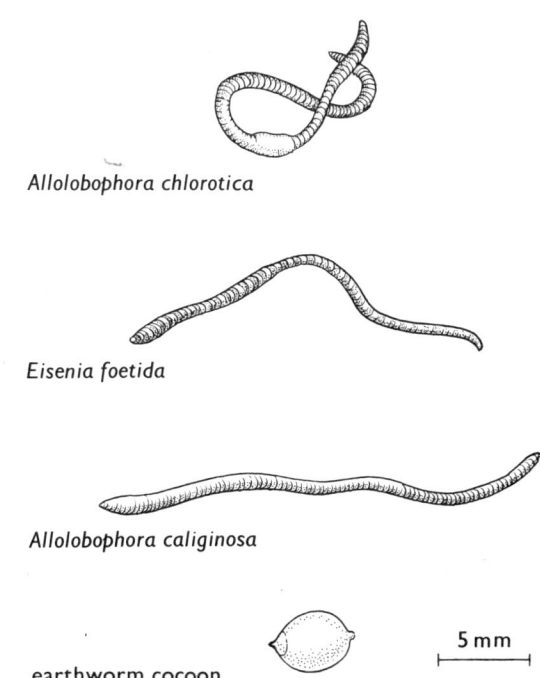

Allolobophora chlorotica

Eisenia foetida

Allolobophora caliginosa

earthworm cocoon

5 mm

This is a fresh water Annelid which spends part of its life as a parasite, and sucks the blood of other animals. There is an anterior sucker which surrounds the mouth, and a posterior sucker. The body is segmented and the worm swims by rippling movements, it is also able to pull itself along surfaces by means of its suckers. Eggs are laid in transparent capsules attached to water weeds.

ARTHROPODA

CLASS CRUSTACEA

10.1.4 The Common Shore Crab (*Carcinus maenas*)

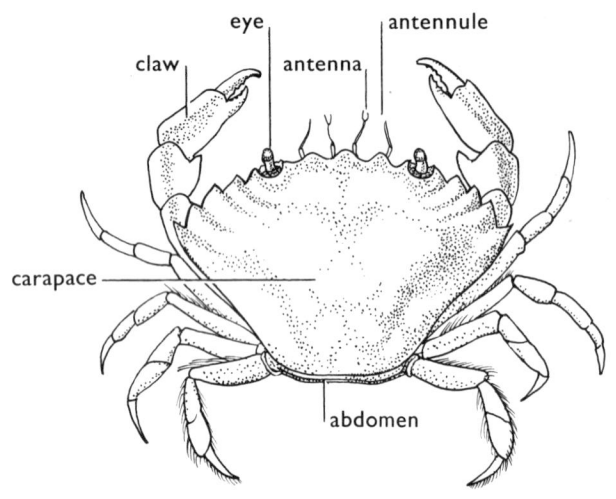

DORSAL VIEW

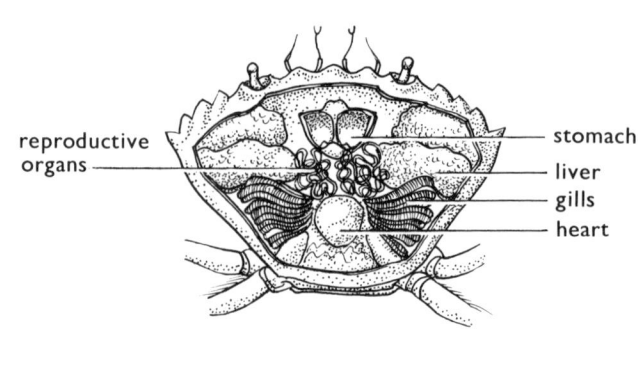

DISSECTION FROM DORSAL SURFACE

The animal inhabits the middle and lower zones of sea shores. The head and thorax are fused to form a large flattened *carapace* and a small abdomen is tucked underneath it. A pair of compound eyes protrude from sockets at the front of the head, and in between them a pair of antennae and a pair of antennules are located. The five pairs of 'walking legs' are robustly constructed, and the crab normally moves sideways on them. The front pair is armed with powerful claws, with which prey is seized. The mouth parts chew it up, and pass it into the mouth. The digestive and respiratory systems are similar to those of the Crayfish (See page 20).

The sexes are separate and after fertilization the eggs are carried by the female attached to the swimmerets under the abdomen. Each egg hatches out into a *zoea* larva which is dispersed in the plankton, and then after further development it moults into a *megalopa* larva (See page 97).

(*a*) Examine live crabs in dishes of sea water. Observe the mouth parts which draw water over the gills. Put a little ink in the water and observe its passage through the gill chamber, emerging at the front end. Place a worm nearby and observe how the crab feeds on it.

(*b*) Pin a dead crab ventral side downwards in a dissecting dish. Remove the dorsal part of the carapace with a stout pair of scissors, and identify the internal organs.

10.1.5 The Acorn Barnacle (*Balanus balanoides*)

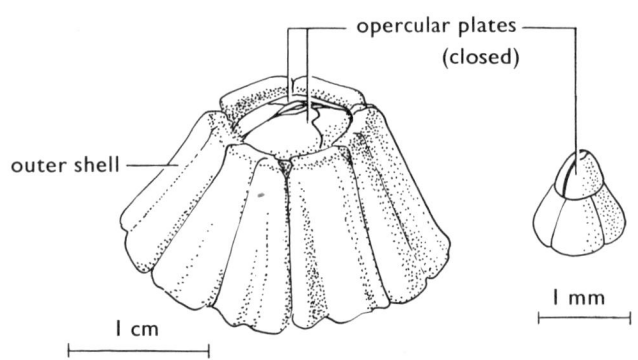

OLD AND YOUNG ANIMAL AT REST

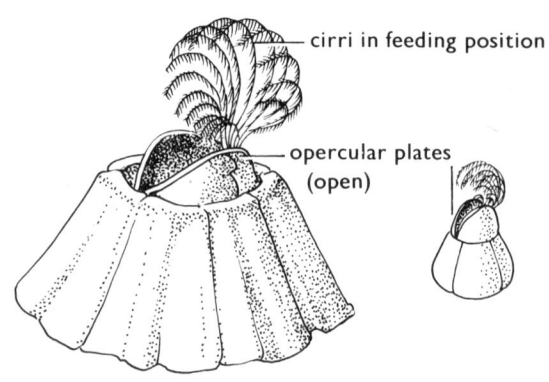

OLD AND YOUNG ANIMAL FEEDING

This sedentary Crustacean is common on rocks between tide marks, often 30,000/m² may be counted. The body is surrounded by hard plates; at periodic intervals the opercular plates open and the feeding appendages (*cirri*) emerge and grasp plankton food which is then drawn into the body. Eggs are shed into the sea, and when fertilized, they develop into *nauplius* larvae which are planktonic feeders. Later a second larval stage (*cypris*) develops, which is bivalved in form, and when fully grown it attaches itself to a hard surface and then grows into the adult barnacle. Frequently young adults can be seen in large numbers covering breakwaters as the tide goes out.

Place a small stone or Mussel covered with barnacles in a dish of sea water, and examine their method of feeding with a hand lens. Observe their behaviour when a shadow is cast over the animals.

10.2 CLASS INSECTA

10.2.6 The Locust
(*Locusta migratoria*)

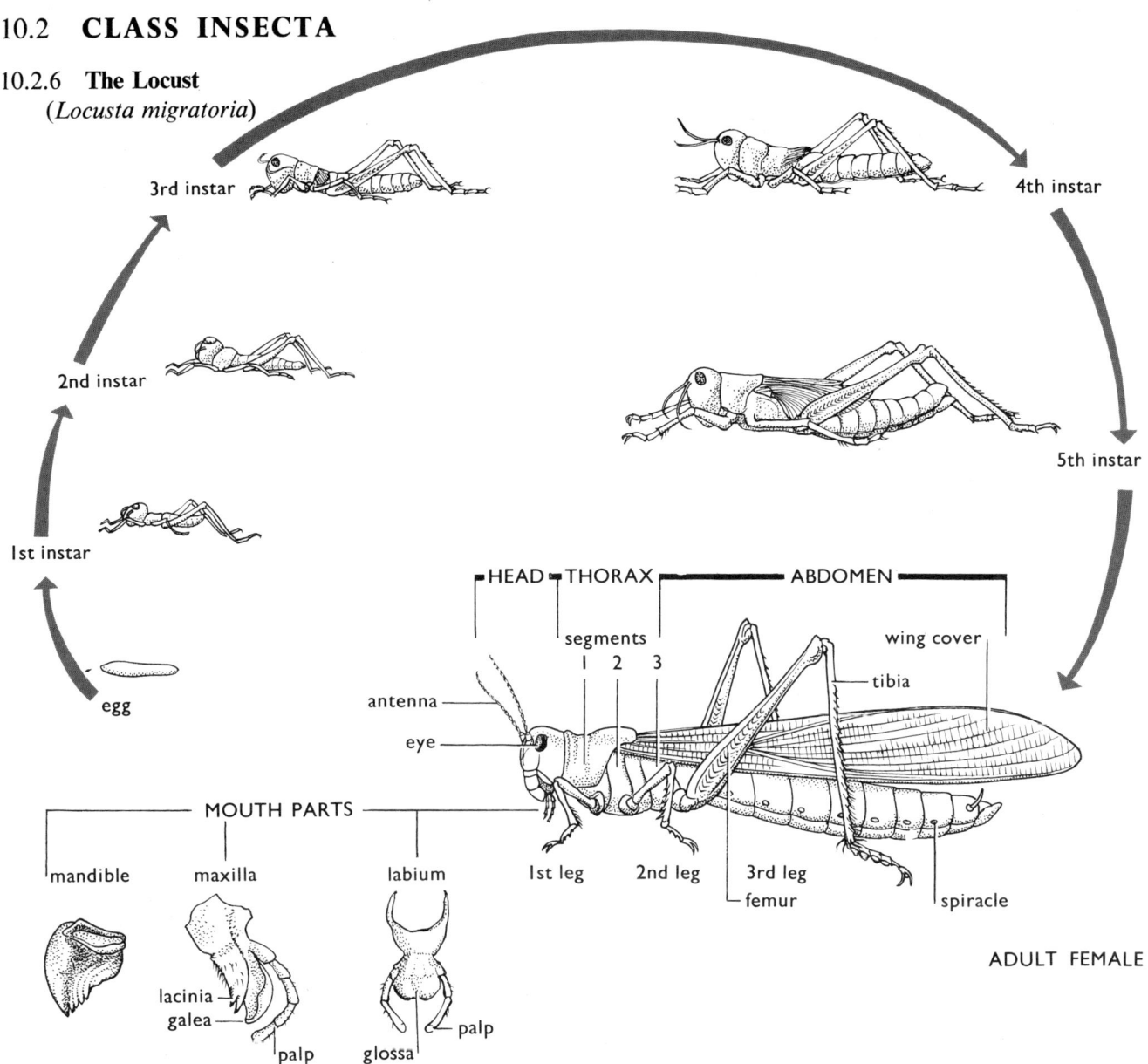

Habitat Tropical Africa, where swarms occur which damage crops. Continuous efforts are made to try to exterminate it. Locusts can be kept in the laboratory in a suitable cage in which its life cycle can be studied.

Structure It is a large insect with a distinct head, thorax and abdomen. The head carries a pair of long segmented antennae, a pair of compound eyes and three pairs of mouth parts—*mandibles*, *maxillae* and *labium*. On the dorsal surface of the thorax there are a pair of wing covers and a pair of wings: on the ventral surface there are three pairs of legs, the posterior pair are the most powerful.

Irritability The antennae and eyes are the main sense organs. Internally there is 'brain' in the head, and a ventral nerve cord inside the body.

Movement The nymphal stages walk and hop actively, the adult is a powerful flier.

Nutrition It is a vegetarian and devours crops voraciously with its strong mandibles.

Respiration A system of trachea opens to the outside of the body by means of spiracles.

Reproduction and life-cycle Shortly after copulation with the male, the female inserts her abdomen deeply into damp sand and lays eggs in pods. The optimum temperature for their development is 35°C and after 11 days they hatch into nymphs. There are five nymphal stages (*instars*) often called hoppers, as they move about actively destroying vegetation on the ground. The adults fly in swarms, and cause great damage to crops when they descend to feed.

Method of examination If a cage is available keep a colony of Locusts under observation. Remove the mouth parts, wings and legs from a freshly killed specimen and study their structure. Dissect the animal and study the digestive and nervous systems in particular. (See also 23.3.3).

10.3 CLASS ARACHNIDA

10.3.1 The House Spider (*Tegenaria atrica*)

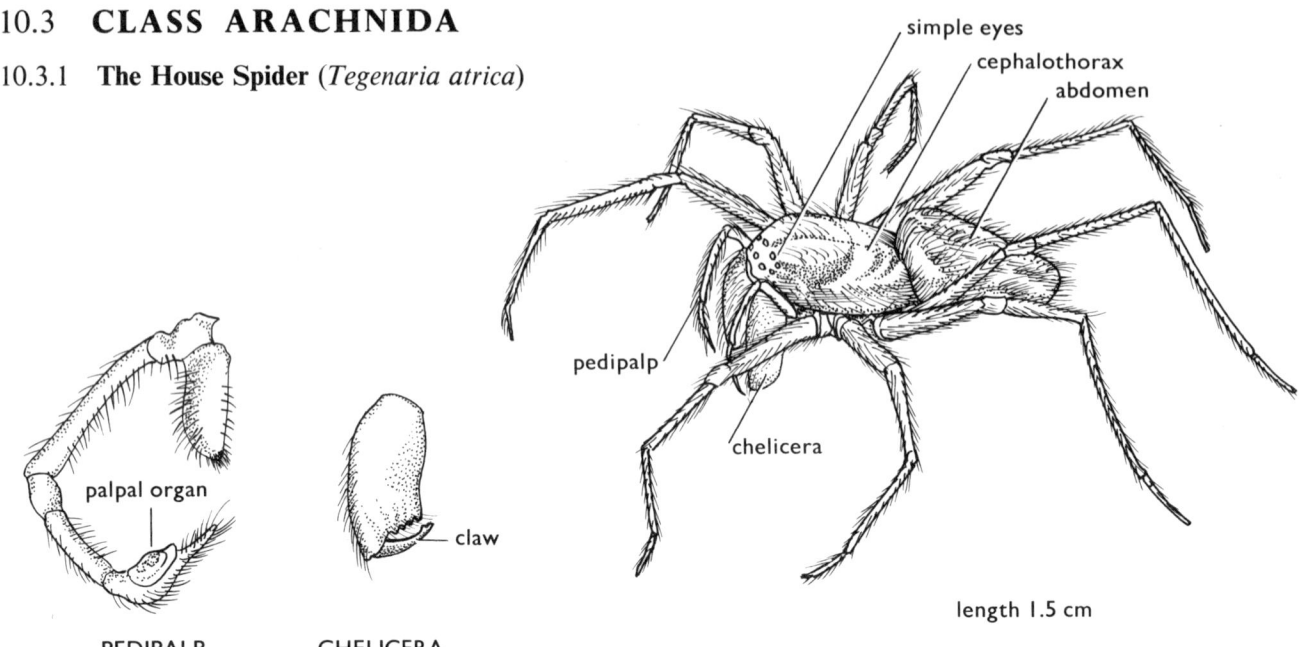

simple eyes
cephalothorax
abdomen
pedipalp
chelicera

length 1.5 cm

palpal organ

claw

PEDIPALP   CHELICERA

Habitat In crevices and corners especially in buildings, where it builds a cobweb of fine non-sticky silken threads.

Structure The horny exoskeleton is divided into cephalo-thorax and abdomen, separated by a slender waist. The head bears 8 simple eyes, a pair of *pedipalps* (sensory) and *chelicerae* (poison jaws). There are 4 pairs of seven-jointed legs arising from the ventral part of the thorax. Spinnerets which form the silk of the cobweb are ventrally placed near the tip of the abdomen.

Irritability and movement There is a 'brain' and a ventral nerve cord. The eyes and *pedipalps* are the main sense organs; the legs are sensitive to touch as well as being used for loco-motion.

Nutrition Prey trapped in the web is poisoned by the *chelicerae*. The mouth is then placed close to it and a protease enzyme is exuded which digests the prey externally. The re-sulting juice is sucked into the spider's gut.

Respiration A pair of lung books opening underneath the abdomen have thin leaves like a book, in which circulating blood is oxygenated. There are also simple tracheae.

Reproduction and life-cycle The male picks up sperms from his reproductive opening and transfers them to the *palpal organ*. After a courtship display, he then uses his *pedipalps* to transfer the sperm to the reproductive opening of the female. Fertilized eggs are laid in cocoons. When the young spiders hatch, they grow by a series of moults.

Method of examination If possible keep a male and female in a vivarium, and examine their behaviour. Examine a living specimen using a hand lens. Study prepared slides of append-ages.

10.4 CLASS MYRIAPODA

10.4.1 The Common Centipede (*Lithobius forficatus*)

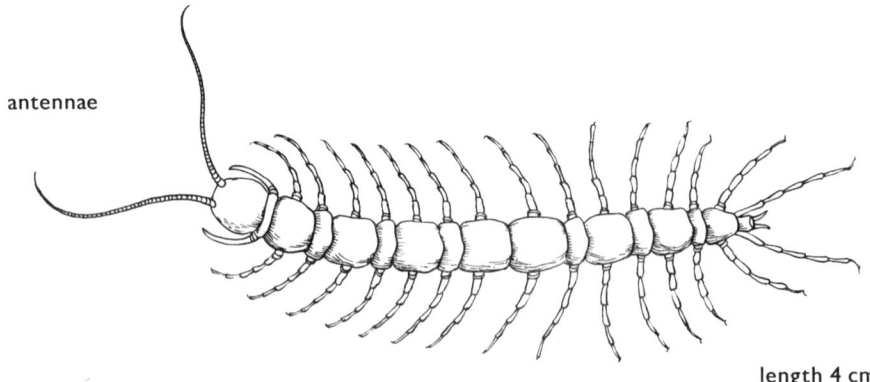

antennae

length 4 cm

It lives in soil and specimens are frequently found under stones and leaves. The head bears a pair of antennae and eye spots. The adult animal has 15 pairs of legs, younger specimens may have fewer. It moves actively, catching its prey by seizing it with poison appendages situated just below the mouth. The female lays eggs singly in the soil.

Keep centipedes in a container of damp soil and leaves and observe their habits.

CHORDATA

SUB PHYLUM UROCHORDATA

13.0.1 CLASS ASCIDACEA

The Sea Squirt (*Ciona intestinalis*)

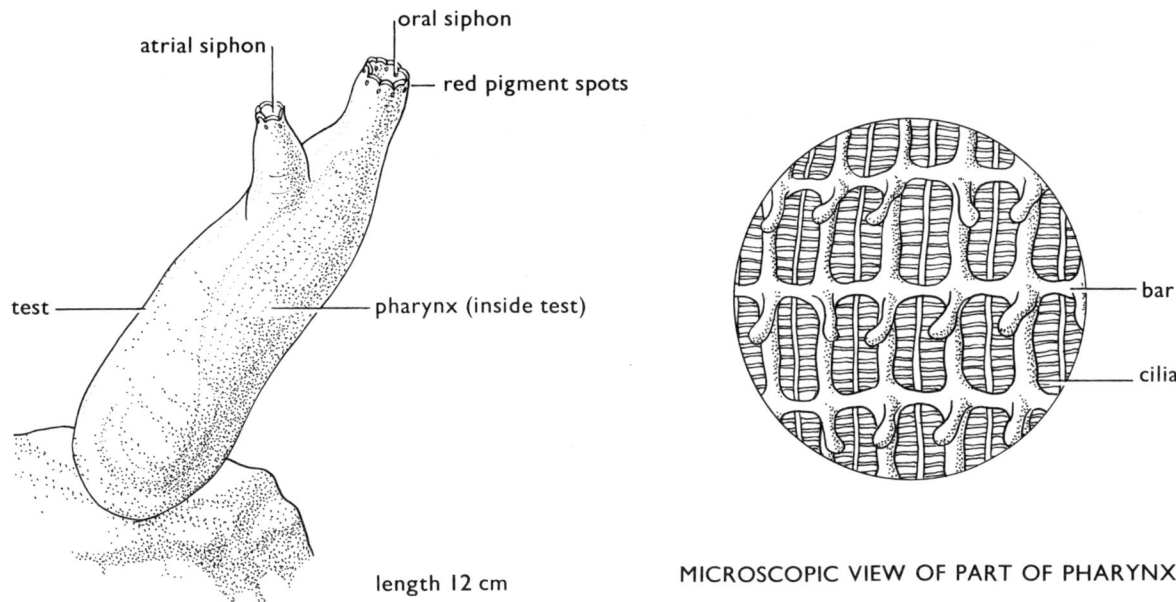

MICROSCOPIC VIEW OF PART OF PHARYNX

This primitive Chordate is a sedentary animal often found on the lower parts of the sea shore attached to rocks, piles, etc. It consists of a stout bag (*test*) penetrated by two siphons surrounded with pigment spots at their tips. Inside, a ciliated meshwork (the *pharynx*) draws sea water in through the oral siphon and filters out food organisms, which are passed into the gut. The suplus water passes out through the atrial siphon. The larval stage is somewhat like a tadpole in appearance.

Observe living specimens in a dish of sea water, noting particularly the current of water they cause, and their ability to contract when touched.

SUB PHYLUM VERTEBRATA

13.0.2 CLASS AGNATHA

The River Lamprey (*Petromyzon fluviatilis*)

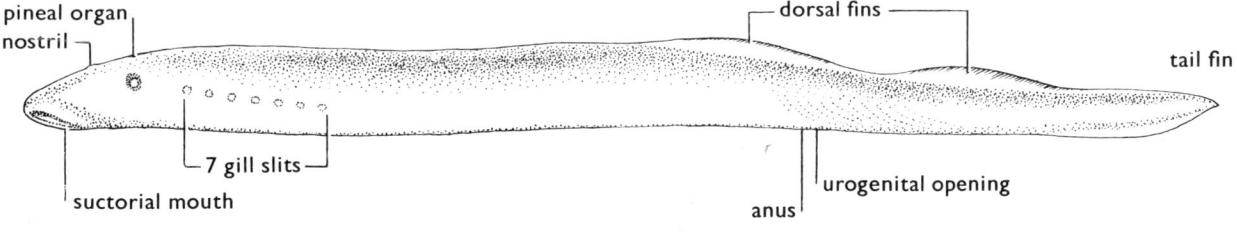

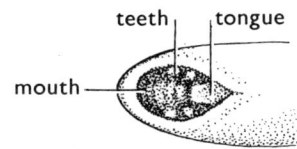

VENTRAL VIEW OF HEAD

It inhabits rivers, the adult reaches a length of about 50 cm. At the head end there is a pair of lateral eyes, a median nostril and pineal organ. There is a simple internal cartilaginous skeleton. The animal swims towards its prey to which it attaches itself with the oral sucker. It rasps the flesh of its host by means of horny teeth, and feeds mainly on blood.

There are 7 pairs of gills which open internally into a respiratory tube. It has a well-developed heart and blood system. Eggs are laid, which hatch and develop into an *Ammocoete* larva, which is a filter feeder, with a pharynx structure similar to that of a Sea Squirt. It lives in the mud at the bottom of the river.

Examine a preserved specimen of an adult Lamprey, and identify the external features, including the structure of the mouth. The arrangement of the internal organs can be examined by slicing the animal with a scalpel from head to tail along its axis in the vertical plane.

ANIMAL PLANKTON

In the Zooplankton every major animal group is represented either in the larval or adult stage. Many larval forms disperse their species horizontally as they wander about in the plankton and then on reaching maturity sink vertically to the bottom. This is especially the case in marine species.

Many larvae are microscopic transparent ciliated organisms and are therefore only easily visible when alive. They can be collected in a plankton net and brought back in a thermos flask for examination in the laboratory. This avoids excessive changes of temperature which would be harmful to the planktonic organisms.

Fresh Water

PROTOZOA

CILIATES

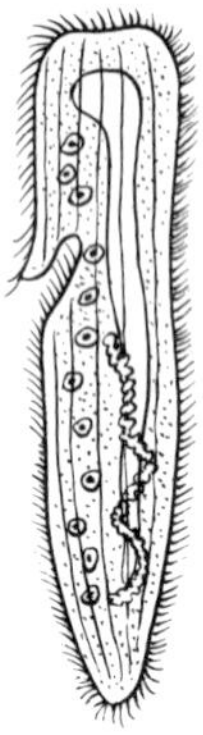

FLAGELLATE

length 2 mm

Spirostomum

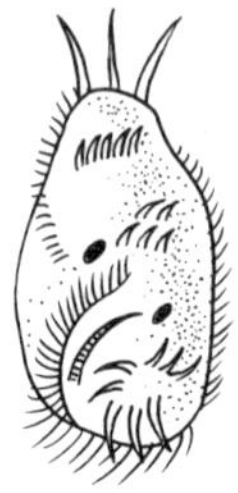

length 100 μ

Stylonichia

length 1 mm

Vorticella

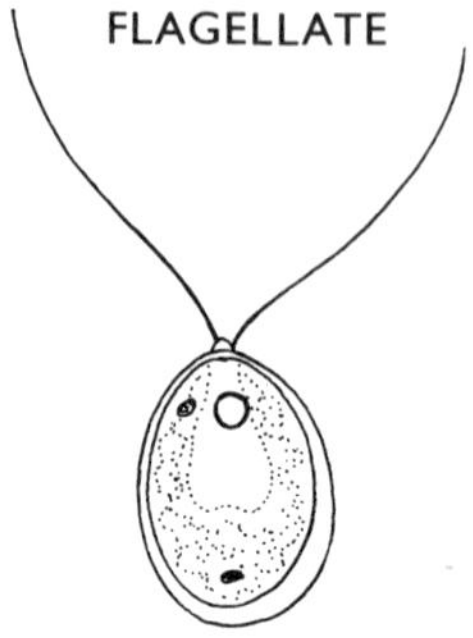

length 10 μ

Chlamydomonas

ROTIFERA

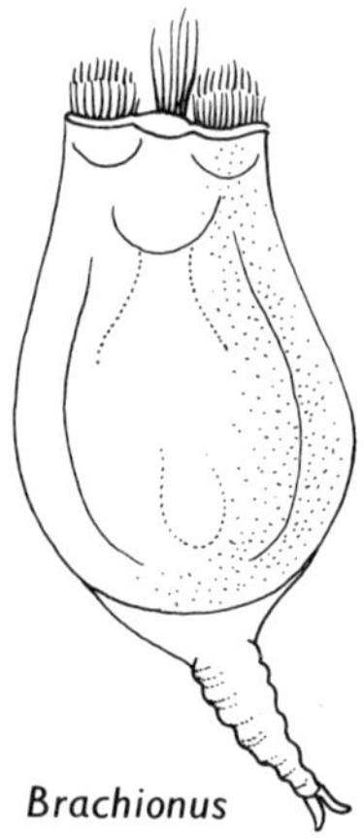

length 1 mm

Brachionus

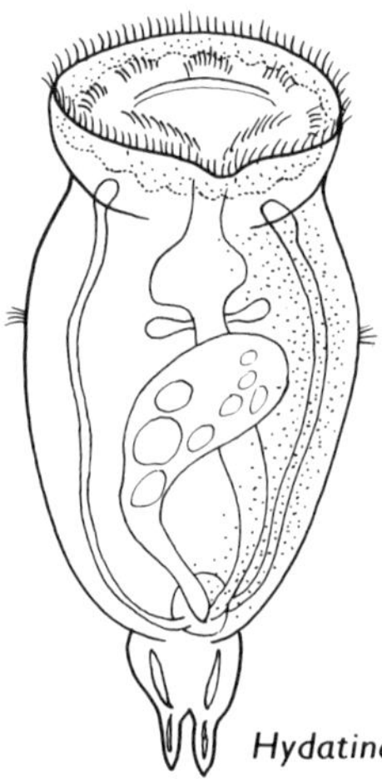

length 1 mm

Hydatina

ARTHROPODA

CRUSTACEA

INSECTA

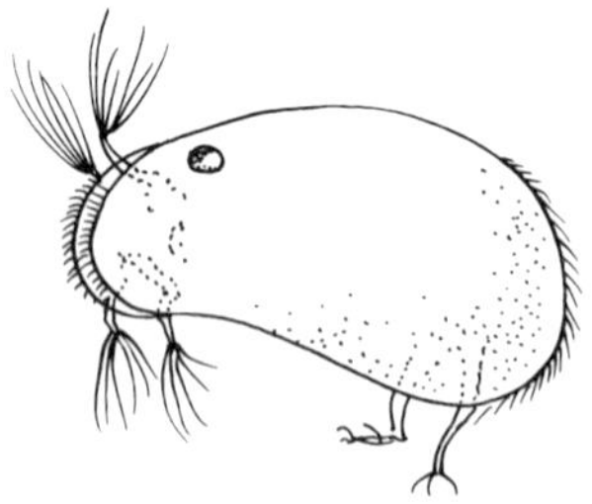

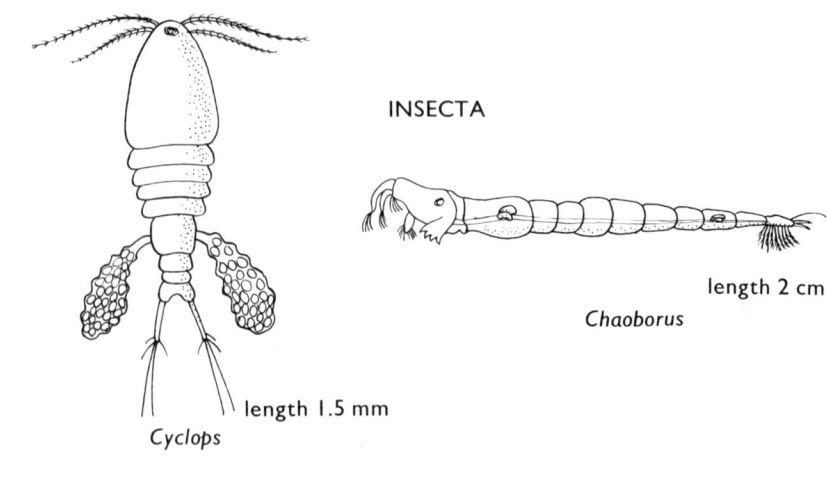

length 1 mm

Cypris

length 1.5 mm

Cyclops

length 2 cm

Chaoborus

Sea Water ANNELIDA Larvae MOLLUSCA

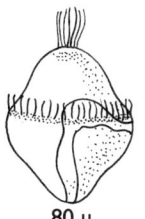

80 μ

Trochophore

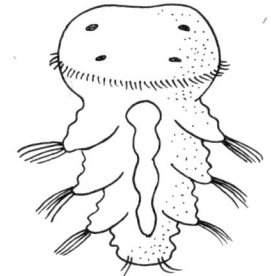

Nectochaeta   length 300 μ

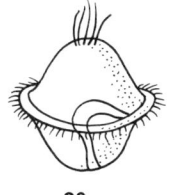

80 μ

Trochophore

Veliger length 200 μ

ARTHROPODA

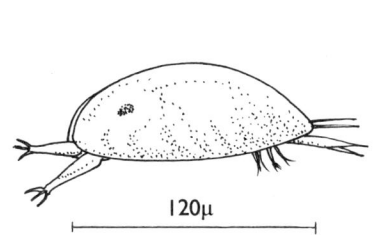

120μ

Cypris (Barnacle larva)

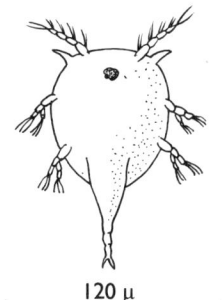

120 μ

Nauplius (Barnacle larva)

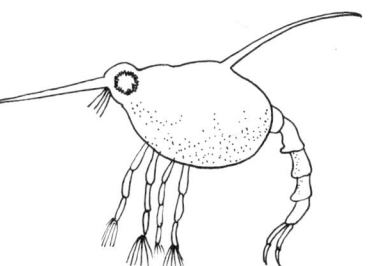

Zoea (Crab larva) length 150 μ

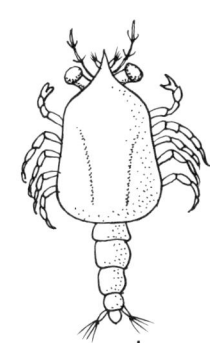

1 mm

Megalopa (Crab larva)

ECHINODERMATA

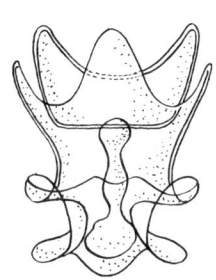

Bipinnaria
(Starfish larva)   80 μ

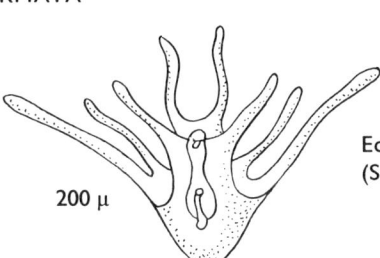

200 μ   Echinopluteus
(Sea Urchin larva)

Adult Animals

CTENOPHORA   CHAETOGNATHA   COELENTERATA

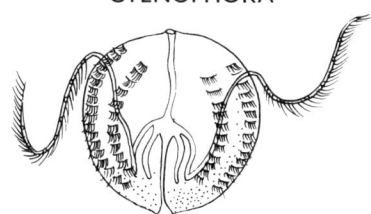

Pleurobrachia 1 cm

Sagitta length 2 cm

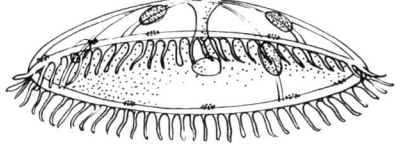

Obelia medusa diameter 1 cm

CRUSTACEA

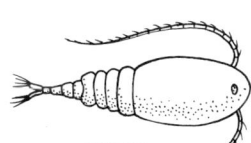

Pseudocalanus length 2 mm

Temora length 2 mm

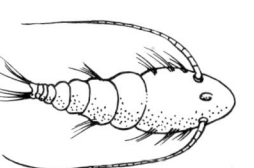

Calanus length 1 cm

97

IV EXPERIMENTS

The purpose of an experiment is to investigate the behaviour of a living organism when various conditions in its environment are changed. Wherever possible an organism is kept under normal conditions, and is the 'control', which is compared with the behaviour of the subject of the experiment.

Experiments can also be made to investigate the chemical and physical properties of living organisms and their products.

18 Experiments on Irritability

Living organisms are sensitive to stimulations produced by changes in their environment. They have sense organs (receptors) of many types, which can detect these changes. Lower organisms have simple receptors, higher organisms have more complicated receptor systems. They make various types of response, which can be listed as follows:

A *tropism* is a response made by a plant organ which grows in a direction determined by the stimulus. A positive response is a growth towards the stimulus, a negative response a growth away from it.

A *taxis* is a movement of a whole organism towards (positive) or away from (negative) a stimulus.

A *nasty* is the response of part of a plant to a change in the intensity of a stimulus surrounding it, e.g. a change in temperature.

A *kinesis* is a change in the rate at which an organism moves due to changes in the strength of the stimulus.

The following experiments can be used to investigate the sensitivity of various organisms.

18.1 LIGHT

When investigating the effect of light, try to keep other factors such as humidity and temperature steady; e.g. be careful that the heat from a lamp does not introduce a temperature effect.

ANIMALS

Experiments on Light from One Direction (Phototaxis)

i. Flatworm (*Planaria* spp.). See 7.1

Place an animal in a small glass dish of water, in as dim a light as possible. Shine a narrow band of light at it from various directions, and investigate the response made by the animal.

ii. Earthworm (*Lumbricus* spp.). See 9.1

Place a worm in a glass tube about 2 cm in diameter and 30 cm long. Cover each end with a cylinder of black paper, each about 20 cm long. Keep the tube steady and horizontal and illuminate the worm with a bright light, moving the two cylinders so as to give a small slit of light at various points along the worm's body. Investigate how different parts of its body react to light.

iii. Water Flea (*Daphnia* spp.). See 10.1.2

Place a number in a small aquarium in a dim light. Shine a bright light on one side of the tank, and note the reaction.

iv. Woodlouse (*Oniscus* spp.). See 10.1.3

Place a strip of damp paper along the length of a wide glass tube approximately 30 cm long. Illuminate one end of the tube and cover the other with a piece of black paper. Introduce 10 animals into the tube, and count the number of individuals at each end at minute intervals over a short period of time.

v. Blow-fly larva (*Calliphora* spp.). Cf. House-fly 10.2.4

Place some young larvae on a piece of damp paper in a dim light. Note the reaction when a narrow band of light is shone on to the larvae.

vi. Minnow (*Phoxinus* spp.). See 13.1.2

Place a number in a small aquarium in a dim light. Shine a bright light on one side and note the reaction.

Experiments on Colour Change

Some aquatic animals change colour according to whether the background is dark or light. Put three glass dishes of water on a dark piece of paper and three others on a white piece. Place one specimen of the following types into each of the different types of background and investigate the nature of the colour change. Crayfish. Minnow. Tadpole.

Investigate the colour change in the adult frog by setting up four experiments. Cover each of four large glass dishes with a piece of glass, and arrange the dishes so that:

(a) is in the light, and has a damp atmosphere;
(b) is in the light, and has a dry atmosphere;
(c) is in the dark, and has a damp atmosphere;
(d) is in the dark, and has a dry atmosphere.

Place a frog in each dish, and notice the colour of the skin before and after half an hour in the experimental dish.

Behaviour of the human eye to light

Get a friend to cover both his eyes with his hands, at the same time keeping them open. Remove one hand quickly and notice the reaction of the pupil of that eye to the sudden exposure to normal light.

Investigate the nature of the blind spot. Cover the left eye, and with the other observe a white card on which there is a black cross and a

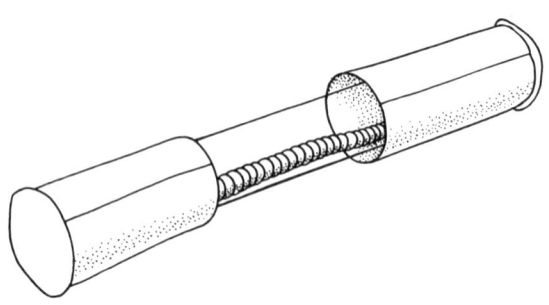

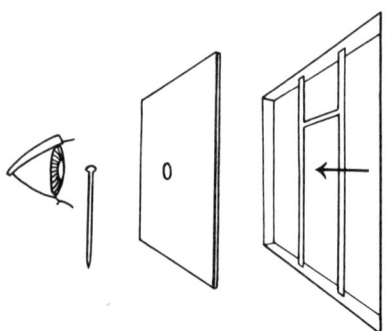

black spot each about $\frac{1}{2}$ cm in size and separated by 6 cm. Concentrate on the spot (have it to your left), and bring the card nearer to the right eye. Note that at a certain point the cross becomes invisible, as it is focused on the blind spot.

Observe the pin-hole camera effect by making a hole with a pin in the centre of a card. Look at a view through the window through the pin-hole and see how much of it is in focus. Remove the card, and observe how the view is seen normally.

Observe that an image on the retina is inverted. Using the same card, hold it about 5 cm in front of one eye. Carefully bring the head of a pin between the card and the eye, observe the outline of the pin in the hole in the card.

Try to obtain a test card, and test the ability to read print of various sizes at the prescribed distances. Similarly use a test card to test the ability to detect colour satisfactorily.

PLANTS

Experiments on Light from One Direction (Phototropism)

i. Pin Mould (*Mucor* spp.). See 14.4.1

Grow a culture of this fungus on a piece of damp bread covered with a cardboard box, with a narrow slit on one side through which light can pass. Investigate the nature of the growth of the sporangiophores.

ii. The Maize (*Zea mais*). See 17.2.3

Grow seedlings of this plant, and mark the coleoptiles with spots of Indian ink 1 mm apart. Expose the seedlings to light from one direction. Note the response of the seedlings compared with those whose tips have been covered with a small silver paper cap. (Seedlings of oats may also be used.)

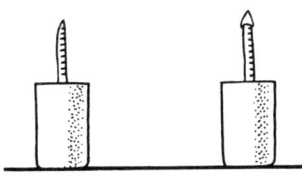

Take six seedlings and mark their coleoptiles with spots of Indian ink 1 mm apart. Place three in a source of light from one direction; one as a control, the other two with a small piece of mica or aluminium foil inserted about 1 mm behind the growing tip, and reaching half way into the coleoptile as shown in the diagram below. Observe the response of these seedlings after 24 hours.

Place the other three seedlings similarly treated, in a dark cupboard, and observe their response after 24 hours.

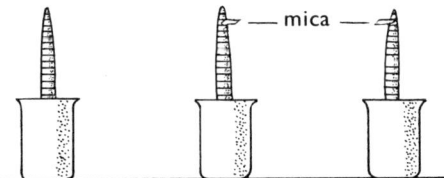
— mica —

iii. Lupin (*Lupinus* spp)

Place a lupin flower spike in a jar of water and illuminate it strongly from one side with an electric lamp for 24 hours. Note the response.

iv. Cress

Support some cress seedlings on a wire mesh placed over a jar of water, so that the roots are in the water. Shine a light from an electric lamp from one side of the jar and note the response in 24 hours.

In all the above experiments set up controls in which the specimens are illuminated with diffused light from *all* sides.

Experiments on Varying Light Intensity (Photonasty)

Place a pot of daisies and a pot of *Oxalis* under a box, so as to keep them dark for several hours. Then expose the plants to normal daylight and observe the response.

18.2 TEMPERATURE

When carrying out observations on the effect of temperature variations on living organisms, try to keep other factors such as light and humidity steady.

ANIMALS

i.(a). Paramecium spp. See 5.2. (Kinesis)

Place a microscope slide containing a culture of these animals in a film of water, so that one end is over a petri dish of warm water, and the other over a dish of cold water. Observe with a hand lens where the *Paramecia* congregate.

i (b). Flatworm (*Planaria* spp.)

Repeat the experiment using flatworms (*Planaria* spp.) (See 7.1.) Place 10 animals in a narrow shallow trough of water, one end over a dish of warm, and the other over a dish of cold water. Count the number at each end at 1 minute intervals over a short period of time.

ii. Woodlouse (*Oniscus* spp.). See 10.1.3 (Kinesis)

Place ten animals in a glass tube, as in 18.1 above. Arrange for one end of the tube to be warm, and the other cool. Count the number at each end at 1 minute intervals, over a short period of time.

iii. Earthworm (*Lumbricus* spp.). See 9.1

Carefully introduce a worm into a narrow glass tube, and close each end with a piece of cotton-wool. Lower the tube into a dish of water at 5°C, and with the aid of a hand lens count the number of pulsations made by the dorsal blood vessel in 1 minute. This can be seen just under the skin on the dorsal surface of the animal. Record the results. Raise the temperature by stages up to 30°C. Make a graph of the readings obtained, plotting temperature against the number of pulsations per minute.

iv. Water Flea (*Daphnia* spp.). See 10.1.2

Place a piece of cotton-wool on the centre of a microscope slide, and make a clearing in the centre of the wool. Moisten it, and with the aid of a pipette place a *Daphnia* in the centre of the wool, so that it is in a kind of nest. Place a split matchstick at each end of the slide, and cover it carefully with another slide, so that the *Daphnia* is held in position in the cotton-wool. Fix the two slides together with a rubber band at each end. Place a deep petri dish containing water at 5°C on the stage of a microscope, and lower the slides containing the *Daphnia* into it. When the animal has settled, focus the heart, which is on the dorsal side of the body, with the low power lens. Count the number of heart beats made in 1 minute. It is best to have a partner to record the times while the observer does the counting. Gradually raise the temperature of the water in steps up to about 25°C, and record the rate of heart beat in each temperature. Make a graph of the results, plotting temperature against the number of heart beats per minute.

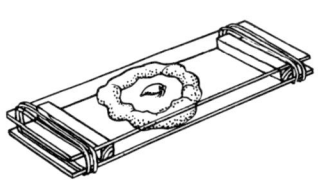

v. Mussel (*Mytilus edulis*). See 11.1

Investigate the rate at which the cilia beat at various temperatures, as observed in a small piece of gill. Open a mussel in a dish of water sea, and cut out a piece of gill about 1 cm square. Place this in a petri dish of sea water at about 5°C, and stand the dish over a piece of graph paper. Place a speck of dust or grain of coloured powder on the edge of the piece of gill, and note the time it takes to travel a distance as measured by the squared paper. Gradually increase the temperature in stages up to about 25°C. Record on a graph the time taken for a speck to travel a given distance against temperature. Be careful not to irritate the gill unduly, as it will emit a mucus which may interfere with the experiment.

vi. Frog (*Rana temporaria*). See 13.2

Sometimes when a frog is dissected, the heart can be massaged so that it starts beating again. If so record the rate of heart beat at different temperatures, and make a graph of the results. The frog should be totally immersed in a dissecting dish containing 3% salt water. Start with the water at 5°C and increase its temperature by 5°C stages, by stirring in warm 3% salt water until a temperature of about 40°C is reached.

Perception of Temperature in Man

Place one knitting needle in hot water and a second one in cold. Blindfold a partner, and map out on his fore-arm which areas detect cold, and which detect hot, by touching the skin at various points with each of the two knitting needles in turn.

Place one hand in a bowl of hot water, and the other in cold water for about 20 seconds. Remove them, and place both hands immediately into a bowl of tepid water. Compare the sensations caused in each hand.

PLANTS

i. Tulips (Thermonasty)

Place a pot of tulips in a cold atmosphere, and later bring the flowers into a warm atmosphere and note the result.

ii. Sensitive Plant (*Mimosa pudica*) (Thermonasty)

Place a lighted match for a few seconds below the terminal leaflets of the plant and note the result.

iii. Pea (*Pisum sativa*). See 17.2.3

Germinate some peas in damp sawdust, so that the roots are about 2 cm long. Mark the roots with Indian ink spots 1mm apart, then place one set in sawdust in a cold temperature, and a second set in a warm temperature for 48 hours. Note the effect of temperature on the rate of growth.

Weigh two separate sets of ten dry peas. Leave one set in cold water, and the other in warm water for 24 hours. Remove them after this time from the water, dry them on blotting paper and weigh again. Compare the amount of water absorbed by each set of peas.

18.3 GRAVITY

When investigating the effect of gravity on living organisms keep light, temperature and humidity steady.

ANIMALS

i. Flatworm (*Planaria* spp.). See 7.1

Place one on a piece of damp glass. Slope the glass slightly and observe the reaction of the *Planaria*. (Geotaxis.)

ii. Crayfish (*Astacus fluviatilis*). See 10.1.1

Place a crayfish in a dish of fresh water and observe its reactions when placed (*a*) on its side, (*b*) upside down.

iii. Woodlouse (*Oniscus* spp.). See 10.1.3

Place ten lice in a wide sloping glass tube, lined with a strip of damp paper. At 1 minute intervals, over a short period of time, count the number of animals at the top and bottom ends of the tube. (Kinesis.)

iv. Cabbage White Caterpillars (*Pieris* spp.). See 10.2.2

Repeat the above experiment, using a batch of ten caterpillars. (Kinesis.)

v. Starfish (*Asterias* spp.). See 12.1

Place a starfish in a dish of sea water and observe its reactions when placed upside down.

vi. Frog (*Rana temporaria*.). See 13.2

Place a frog in a large empty aquarium tank covered to prevent the animal from jumping out of it. Gently tip the tank in different directions and observe the posture of the frog's head. If possible, place the tank on a revolving turntable, and note the position of the frog's head as the animal is revolved.

PLANTS

These experiments should be carried out in the dark.

i. Pin mould (*Mucor* spp.). See 14.4.1

Grow a specimen of this fungus on a piece of damp bread. When the sporangiophores emerge, slope the piece of bread at about 45° and investigate their reaction after 24 hours. (Geotropism.)

Experiments on Roots (Positive geotropism)

ii. Broad Bean (*Vicia faba*). See 17.2.3

Germinate some beans until the roots are about 2 cm long. Make marks on them about 1 mm apart with Indian ink. Place the beans in different positions by pinning them to a piece of cork covered with a thin layer of cotton-wool. Stand the cork vertically in a dish of shallow water. Observe the growth of the roots after 24 and 48 hours. Also mount beans on the cork, from which the terminal 2 mm of the roots have been removed, and note the response.

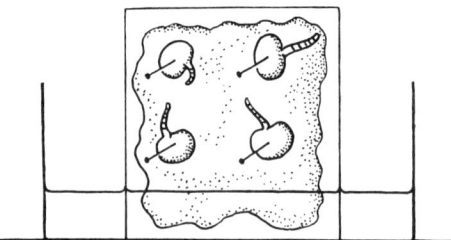

In order to counteract the influence of gravity place some beans with roots approximately 2 cm long on a slowly revolving klinostat. Observe the resultant growth after 24 and 48 hours.

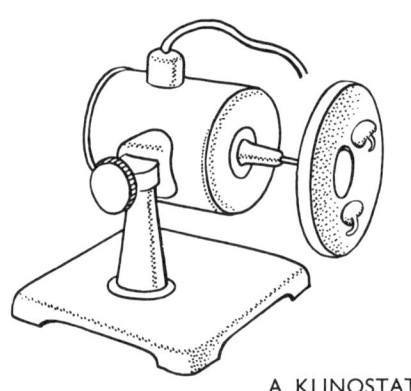

A KLINOSTAT

Increase the stimulus on the roots by using centrifugal force. Germinate some beans in a tin of sawdust and then revolve the whole tin rapidly for 48 hours on a horizontal wheel driven by an electric motor. At the end of this time observe the direction of the growth of the roots in each plant.

Experiments on Stems (Negative geotropism)

iii. Maize (*Zea mais*). See 17.2.3

Germinate some maize seeds in specimen tubes of sawdust. When the coleoptiles emerge, make Indian ink marks 1 mm apart on them. Place the tubes on their sides, and observe the resultant growth of the coleoptiles after 24 and 48 hours. Repeat the experiment with some seedlings from which the terminal 2 mm of coleoptile has been removed. (Seedlings of oats may also be used.)

iv. Lupin (*Lupinus* spp.)

Place a lupin flower spike in a jar of water, by inserting it through a hole in a tight-fitting cork. Make this hole watertight with a little vaseline. Turn the spike on its side and observe the result in 24 hours.

v. Geranium (*Pelargonium* spp.)

Cut a slit in a piece of square plywood, and cover the soil in a pot containing a geranium with it. The stem of the plant should be inserted in the slit. Turn the plant upside down in a dark cupboard, and observe its reaction after 24 and 48 hours.

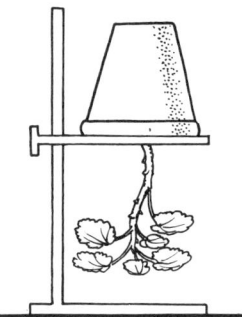

18.4 WATER

ANIMALS

i. Flatworm (*Planaria* spp.). See 7.1

Make a shallow trough, along which water can flow in a slow stream from a tap. Place specimens of *Planaria* in the moving water, and note their reactions to its movement. (Rheotaxis.)

ii. Earthworm (*Lumbricus* spp.). See 9.1

Place an earthworm in a long, wide glass tube dry at one end and with damp cotton-wool at the other end. Cork the tube up at each end, and observe the reaction of the worm. (Kinesis.)

iii. Crayfish (*Astacus fluviatilis*). See 10.1.1

Place a crayfish in a shallow trough of moving water, and observe its reactions. (Rheotaxis.)

iv. Woodlouse (*Oniscus* spp.). See 10.1.3

Cover a shallow trough with a piece of glass. Inside it support a strip of perforated zinc, so that it is over a dish of water at one end, and a dish of $CaCl_2$ at the other. Introduce ten woodlice, and count the number of animals at each end at 1 minute intervals over a short period of time. (Kinesis.)

perforated zinc trough covered with a sheet of glass

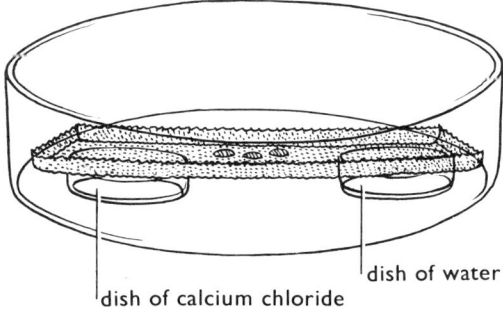

dish of water

dish of calcium chloride

PLANTS

i. Common Moss (*Funaria* spp.). See 15.1

Mount a spore capsule from which the operculum has been removed on a dry slide, and observe it under the low power of the microscope. Breathe on it gently, and observe the movement of the peristome as moist air reaches them.

ii. Fern (*Dryopteris* spp.). See 16.1

Mount an old sorus from which the indusium has withered on a dry slide, and observe it under the low power of the microscope. Breathe on it gently and observe movement in the annulus of a sporangium.

---

iii. Pine (*Pinus* spp.). See 17.1.1

Place mature pine cones alternately in a dry and damp atmosphere and observe their behaviour.

iv. Flowering plants

Place a porous pot (or a plant pot with its drain hole sealed with a cork) containing water in the centre of a seed-box containing dry soil. Place some germinating seeds in the zone around the porous pot. At 24-hour intervals uncover some of the seedlings, and observe the reaction of their roots to water seeping out of the pot. (Hydrotropism.)

18.5 CHEMICALS

ANIMALS

i. *Amoeba* spp. See 5.1. *Paramecium* spp. See 5.2

Mount some specimens of *Amoeba* in a drop of water on a slide, and cover with a coverslip. Introduce a weak solution of sodium bicarbonate at one edge of the coverslip from a fine pipette; i.e. weak carbonic acid solution. Observe the reactions of the animals. (Chemotaxis.)

Repeat this experiment using *Paramecium*.

Repeat both experiments using a very weak solution of acetic acid. In the case of *Paramecium* note the reaction of the trichocysts.

ii. *Hydra* spp. See 6.1

Place an animal in a cavity slide, and cover it with a coverslip. Introduce a very weak solution of acetic acid from a fine pipette, and observe the action of the stinging cells, examining the tentacles under both low and high power.

iii. Earthworm (*Lumbricus* spp.). See 9.1

Place an earthworm on a piece of damp paper. Hold a piece of cotton-wool soaked in ammonia solution near to the worm, and estimate the distance at which it can detect this chemical.

iv. Blow-fly larva (*Calliphora* spp.). Cf. House-fly. See 10.2.4

Place a few larvae on a piece of damp paper, and a piece of meat at one end. Observe their reactions. (Chemotaxis.)

v. Sensitivity in man

Blindfold a partner, and assess his ability to recognize various smells, by holding small bottles containing common smelling compounds near his nose.

Make up solutions of salt, sugar, quinine and vinegar. Touch a small drop of each of these solutions in turn from a clean glass rod on to different parts of your partner's tongue. Map out the areas sensitive to the different substances. Repeat these experiments with the nose clipped, so that the sense of smell is eliminated.

Make up a solution of 1 g sugar in 100 cm^3 distilled water. From this make up small quantities of more dilute solutions. Find out the strength of the weakest solution which your partner can detect. As control experiments, occasionally touch the tongue with distilled water only.

PLANTS

i. Pin Mould (*Mucor* spp.). See 14.4.1

Place a small portion of mycelium from a flourishing growth of this fungus on to agar jelly contained in a petri dish. Place a piece of bread on the jelly near the fungus, and note its reaction. (Chemotropism.)

ii. Flowering plants

Bring a piece of cotton-wool soaked in ammonia solution close to the terminal leaflets of a sensitive plant (*Mimosa pudica*), and observe the reaction. (Chemonasty.)

Place a small piece of hard white of egg, or a small insect, on the tentacles of a sundew (*Drosera* spp.) and observe the reactions with a hand lens. (Chemotropism.)

Warm a little agar jelly so that it just melts, and put a drop on a microscope slide. Into this put a mixture of pollen grains and shredded stigma from a flower. Specimens of lupin, pea or hyacinth are suitable. Place the slide in a moist atmosphere in the dark for a few hours. Later observe it under the microscope, and examine the growth of pollen tubes from the pollen grains. (Chemotropism.)

18.6 TOUCH

ANIMALS

i. *Paramecium* spp. See 5.2

Observe the reactions of *Paramecium* when it bumps into small obstacles such as water plants, when examined under the low power of the microscope.

ii. *Hydra* spp. See 6.1

Observe a specimen in a drop of water on a cavity slide. Touch the tentacles gently with the point of a seeker, and observe the reactions with a hand lens.

iii. Anemone (*Actinia* spp.). See 6.2

Observe an anemone in a small dish of sea water. When its tentacles are fully extended touch one gently with a seeker, and observe the response. Progressively stimulate the anemone more strongly and observe its reactions. Observe the behaviour of the animal when it is fed on small pieces of chopped earthworm or meat.

Touch various parts of the following animals with the point of a seeker, and try to determine where they are most sensitive to the stimulus of touch. Earthworm, Crayfish, Cockroach, Caterpillar.

iv. Woodlouse (*Oniscus* spp.). See 10.1.3

Place ten woodlice in a small glass dish, one-half covered with sand, and the other with smooth paper. Count the number of animals on the smooth and rough surfaces at 1 minute intervals over a short period of time. (Kinesis.)

v. Mussel (*Mytilus edulis*). See 11.1

Open the two shells of a mussel, and place it in a dish of sea water. Notice the reaction of the foot and palps when they are touched gently with a seeker.

vi. Starfish (*Asterias* spp.). See 12.1

Observe a starfish in a small dish of sea water. Touch the tube feet on various parts of the body and also the terminal tentacles on each arm, and observe the response.

vii. Sensitivity in man

Carry out the following experiments on a blindfolded partner. Find the smallest piece of paper which can be detected when dropped from a height of a few cm on to both surfaces of a hand, and on to the forehead. Repeat the experiment several times, and take the result for which the answer is given correctly most times.

Set the points of a pair of dividers to various distances and discover the smallest distance at which two points can be recognized on various parts of the skin.

PLANTS

i. Pea (*Pisum sativa*). See page 84

Touch one side of a tendril at regular intervals with the point of a seeker. Determine how long it is before a response is made. (Thigmotropism.)

ii. *Oxalis acetosella*

Tap the leaflets gently with a pencil for about 2 minutes. Observe the reaction of the plant. (Thigmonasty.)

iii. Sensitive plant (*Mimosa pudica*)

Touch the tips of the leaflets with a pencil, and observe the response to stimuli of various strengths. (Thigmonasty.)

18.7 SOUND

Hold a meter rule near to a partner's ear in a quiet room. The other ear should be plugged. Move a quietly ticking watch outwards along the rule, and record the distance at which it is no longer heard. Bring the watch back towards your partner again, and record the distance at which it is heard. Take the mean of the two readings. Repeat the experiment with the other ear and observe whether the sensitivity of the two ears is similar.

Plug both your ears with cotton-wool, so that you cannot hear a watch close to. Touch the watch against the teeth, forehead, etc., and observe whether the tick can be heard by conduction through the bones.

Set a tuning fork humming, or produce a continuous note from a musical instrument. Plug each ear alternately in quick succession with a finger. Observe whether the pitch of the note sounds the same when detected by each ear.

Obtain a Galton Whistle, or oscillator, and observe the highest note which can be detected by each ear.

Obtain a suitable cathode-ray oscilloscope, amplifier and sensitive microphone. Study the nature of the curves obtained with this apparatus and displayed on the oscilloscope screen, when various sounds are made into the microphone.

19 Experiments on Movement

Observe the various methods of movement found in living organisms.

19.1 ANIMALS

AQUATIC ANIMALS

Ciliated movement Observe this in *Paramecium*. *Planaria*. *Fasciola* miracidia larvae. Mussel gills, Earthworm nephridia. Lining of the buccal cavity of a frog.

Appendages Crayfish. Water flea, Gnat larvae. Fresh water shrimp (*Gammarus* spp.). Observe this animal on a cavity slide in a drop of water under the low power of the microscope. Note the movements of the muscles within the exoskeleton near a joint.

Pseudopodial movements *Amoeba*.

Gliding Observe the movement of the following animals on the side of a glass tank in water. *Hydra*. Pond snail. Sea anemone.

Tube feet Observe the method of movement of a starfish.

Tails Minnow. Tadpole.

TERRESTRIAL ANIMALS

Earthworm Observe animals moving in their tubes in a glass-sided wormery. Place an earthworm on a piece of damp paper, and observe its method of movement.

Insects Observe the method of movement of blow fly larvae, and cockroaches.

Frog. Rat Observe these animals alive. Also study the nature of the skeleton and the structure of the joints in preserved specimens.

AERIAL ANIMALS

Observe the structure of insect wings. See 10.2.1, 10.2.2, 10.2.3, 10.2.4 and 10.2.6.

Study the structure of the bird's wing and feathers. See 13.4.1.

19.2 PLANTS

Autonomic movements

These are natural movements which occur in plants while they are growing, and are not influenced by external stimulations. Observe the growth of pea tendrils, and the twining habit of a convolvulus.

Nutation

Many stem and root apices spiral as they grow. To observe this, place a maize seedling in a small pot on the base of a retort stand. Clamp two pieces of glass one above the other in line over the plant. Place a mark on the upper piece of glass, look at the tip of the plant in line with the mark. Put a piece of sticky paper on the lower piece of glass in line with the mark and the stem tip, and repeat this at intervals of several hours. In this way record the movements of the stem tip.

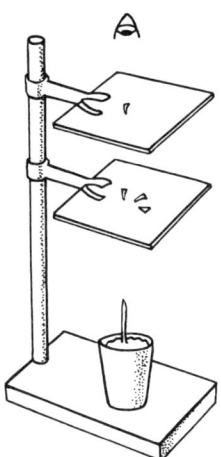

Tactic movements

Observe the movements of plant male gametes, using *Fucus* (see 14.2), Fern prothallus (see 16.1) and *Saprolegnia* (see 14.4.2).

Protoplasmic movements

Mount stamen hairs of *Tradescantia* (See 17.2.9) or cells of *Elodea* in water on a slide. Observing the streaming movements of the cytoplasm under the low and high power of the microscope.

Turgor movements

Strip the epidermis off an iris leaf, and mount a small portion in water on a slide and cover with a coverslip. Observe the position of the guard cells in the stomata. Now introduce a strong salt solution under the coverslip from a pipette, and observe the movements of the guard cells.

Mount a thin section of a piece of beetroot on a slide in water, and cover with a coverslip. Observe the distribution of the red pigment in the cells. Introduce a strong salt solution and observe the movements of the pigment.

Similarly, using the stamen hairs of *Tradescantia* or the stigma hairs of a Geranium, observe the effect of mounting the specimens in water, and then replacing it by a strong salt solution.

20 Experiments in Nutrition

20.1 THE NATURE OF FOOD

Food consists of (*a*) high energy carbon compounds consisting of large complex molecules. The energy contained in these compounds is released in living organisms at comparatively low temperatures in their bodies, by the action of organic catalysts called enzymes. The compounds can also be broken down when heated experimentally.

Fuel food+Oxygen = Carbon dioxide+Water+Energy

(*b*) There are also non-fuel substances in food—minerals, vitamins and water.

20.1.1 Experiments to Demonstrate the Nature of Food

(*a*) To show that fuel foods break down to carbon dioxide and water. Heat a few grams of sugar in a test tube, and allow the escaping fumes to pass from a bent delivery tube on to a piece of blue cobalt chloride paper. The paper will turn pink if water is present in the fumes. Allow the fumes to pass through clear lime water, which will turn milky if carbon dioxide is present.

(*b*) Place some grass in an evaporating basin, and heat it strongly so that it is completely burned to carbon dioxide and water. Note whether any mineral ash remains after the combustion.

20.1.2 Types of Food Substances, and Experiments to Identify Them

HIGH ENERGY FUEL FOODS

Carbohydrates

Cellulose. Add a little chlor-zinc-iodide to a small piece of cotton-wool in an evaporating dish. A blue colour denotes the presence of cellulose.

Starch. Scrape a few potato grains into the bottom of a test tube, and half fill it with water. Warm the water to break up the grains, then cool the test tube. Add a few drops of iodine and observe the purple colour which develops, a test for starch. Warm the test tube slowly in a water bath and using a thermometer observe the temperature at which the purple colour disappears. Allow the test tube to cool again and note the temperature at which the colour reappears. So determine the temperature below which this test for starch can only be carried out reliably. Experiment with different strengths of starch solution to determine how sensitive the test is. Mount a few starch grains on a slide in a drop of water under a coverslip. Using a small piece of filter paper draw the water away and replace it with iodine (See 2.6, page 6). Observe the colour change under the low power of the microscope.

Glucose. Run 5 cm³ of Benedict's solution into a test tube then drop in a few grains of the sugar previously picked up on the point of a seeker or mounted needle. Heat the test tube by placing it in a water bath, and raise the temperature to boiling point. Observe the red precipitate which develops, which is test for glucose. Carry out experiments with solutions of varying strengths to discover how sensitive this test is.

Sucrose. Benedict's test will not work directly on sucrose. The sucrose can be converted into glucose by boiling with a few drops of dilute hydrochloric acid for 1 minute, then the test can be applied.

Maltose. Carry out the test for this sugar in the same manner as that for sucrose.

(Benedict's solution contains copper sulphate, sodium citrate and

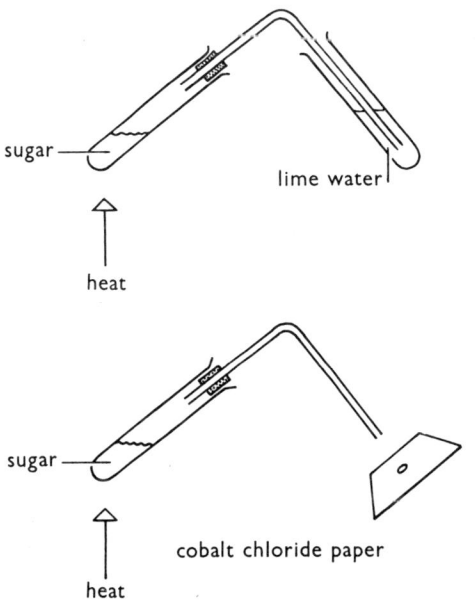

103

sodium carbonate. The red precipitate formed in the test is cuprous oxide.)

Fats

Place a small quantity of margarine in an evaporating basin and heat until it melts. Measure the temperature at which this occurs with a thermometer. Continue the heating and observe the burning of the fat. This is a simple test for fats.

Rub a small piece of margarine on to the centre of a piece of filter paper, and hold it up to the light. Note the translucent grease spot which is made on the paper. Now dissolve the fat off the paper with benzene and allow it to dry. Note whether the fat is removed by the benzene.

Compare the appearance of the translucent spot with a spot of water placed on the same piece of filter paper. Note the effect of warming a piece of filter paper gently, having both a translucent spot of fat, and a spot of water on it.

Place a small amount of margarine in the bottom of a test tube, cover with about half a test tube of water, and heat gently. Notice that the fat breaks into small particles to form an emulsion. Add a little Sudan III stain to this emulsion, shake and allow to settle. The stain turns the fat particles red.

Proteins

Heat a small portion of egg white in an evaporating basin and note the charring and burning.

Xanthoproteic test

Place about 5 cm³ of egg white on the bottom of a test tube and carefully add a few cm³ of concentrated nitric acid. Warm the mixture gently until a yellow colour develops. Cool, and then add a little ammonium hydroxide solution carefully a drop at a time. An orange colour results.

Millon's test

Add a few drops of Millon's reagent to a few cm³ of egg white in a test tube. Heat the test tube gently, and observe the red colour which develops.

Biuret test

Place a few cm³ of egg white in the bottom of a test tube. Add a few cm³ of 5% sodium hydroxide solution and a drop of 1% copper sulphate solution. A violet colour results.

Other tests

Place a few cm³ of egg albumen in a hard glass test tube, add soda lime and heat. Ammonia is evolved.

Boil a few cm³ of egg albumen with 40% sodium hydroxide solution for 3 minutes. Add a little lead acetate solution. A black precipitate develops indicating the presence of sulphur in the protein.

Confirm that all the above tests are specific, that is to say that protein tests do not operate on fats and carbohydrates, or carbohydrate tests on fats and proteins.

NON-FUEL FOOD SUBSTANCES

Minerals

Chop cabbage leaves into fine pieces and heat strongly in a large evaporating basin until only ash remains. Using standard chemical tests, identify the presence of the following substances in the ash.

Pick up a little of the ash on a piece of platinum wire, and identify metals by the colours produced when the ash is held in a bunsen flame. Sodium—yellow. Calcium—red. Potassium—violet, when viewed through a piece of cobalt glass.

Iron. Dissolve a little of the ash in dilute nitric acid, and add a little potassium ferrocyanide solution. A blue coloration is a test for iron.
Chloride. Dissolve a little ash in dilute nitric acid. Add a little silver nitrate solution. A white precipitate is a test for chloride.
Phosphate. Dissolve a little ash in dilute nitric acid and add some ammonium molybdate solution. A yellow coloration is a test for phosphate.
Sulphate. Dissolve a little ash in concentrated hydrochloric acid. Dilute the solution and filter it into a clean test tube. Add barium chloride, a white cloudiness is a test for sulphate.
Nitrate. Dissolve a little ash in water. Add a little 0·5% solution of diphenylamine in concentrated sulphuric acid. A blue coloration is a test for nitrate.

Vitamins

There are no easy tests for all the vitamins with the exception of Vitamin C (Ascorbic acid) which is a reducing agent.

Test for Vitamin C

A solution of the dye dichlorophenolindophenol (DCPIP) is decolorized by reducing agents such as Vitamin C. Place some DCPIP solution in a small syringe and measure out 1 cm³ into a test tube standing in a test tube rack. Place some fruit juice such as lemon or orange into a second syringe, and measure out the volume of juice required to decolorize the DCPIP. If necessary stir the mixture in the test tube with a clean glass rod, but do not shake it at any stage.

Compare the strength of the Vitamin C solution in the fruit juice by repeating the experiment using a solution of ascorbic acid of known strength.

20.1.3 Tests for Various Substances in Common Foods

Apply the above tests systematically to various common foods and identify the occurrence of fuel foods, minerals and Vitamin C in them. Always use small amounts of food materials, and carry out the experiments on a small scale. Investigate the content of bread, potato, apple, vegetables, milk, butter, margarine, egg, cheese, meat, chocolate and fruit juices.

20.1.4 Digestion of Fuel Foods

Before these substances can be absorbed in the tissues of an animal, they have to be broken down into simpler soluble materials, and this is carried out in the process of digestion, by enzymes. Enzymes are catalytic substances made by living cells, which speed up chemical reactions in living organisms. They do this at temperatures of 0°–40°C and in alkaline, neutral or acid conditions according to the type of enzyme. Enzymes are specific, i.e. they take part in definite actions, and they are named by adding the suffix '–ase' to the type of substrate on which they act.

Carbohydrases convert starches to glucose.
Lipases convert fats to fatty acids and glycerol.
Proteases convert proteins to amino-acids.

EXPERIMENTS ON ENZYME ACTIONS

Carbohydrases

Make up a weak starch solution in a test tube, and place it in a water bath at 37°C (body temperature). Transfer half of this to another test tube and test whether there is any sugar present using Benedict's test. To the other half add a few drops of a weak solution of diastase. Place a series of spots of iodine on a white tile, and remove a drop of the starch solution at 1 minute intervals, mixing each drop with one of the spots of iodine in succession. So long as there is any starch present, a purple coloration will result, but when it has all been digested, iodine will be unaffected.

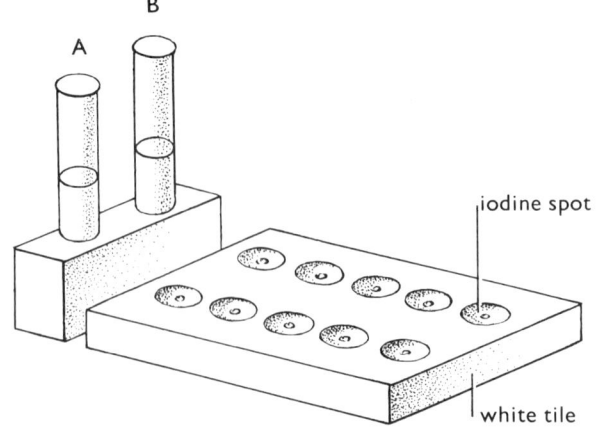

Carry out a similar experiment using saliva instead of diastase. Place three clean test tubes in a beaker of warm water kept at 37°C. Into two of them place about 10 cm³ of starch solution and into the third some saliva made up to about 10 cm³ with distilled water. Using a clean dry white tile, spot out two rows each of five drops of iodine. Add 1 cm³ of the saliva solution to test tube A noting the time and stir in immediately with a clean glass rod. At intervals of 15 seconds afterwards, place a spot of the mixture on successive spots of iodine on the tile, cleaning the glass rod after each test. If there is still any starch present the iodine spot will turn purple. Record the time of the iodine spot whose colour is not changed, and this will give an indication of the speed of starch digestion by the saliva. Now repeat the experiment using the starch solution in test tube B as a control, by adding 1 cm³ of distilled water to it, instead of saliva. Test the solution remaining in test tube A with the Benedict's test to discover whether there is any sugar present. It will be necessary to heat it first with dilute hydrochloric acid, to convert any maltose present into glucose, as Benedict's test will not give a direct result on maltose sugar.

These experiments may be repeated at different temperatures, and the rate of digestion of the starch solution should be recorded for each temperature chosen.

Proteases

Take three test tubes, into each place a small piece of the hard-boiled white of an egg. Cover with water and add one drop of dilute hydrochloric acid to test tube 1 ; one drop of dilute sodium hydroxide solution to test tube 2 ; leave the third test tube neutral. To each add a little pepsin (a protease), and leave in a warm place for several hours. Compare the rate at which the egg white is digested in each of the three test tubes. Set up a control experiment to which no pepsin has been added. (One small crystal of thymol placed in each test tube will prevent the egg white from decaying if the experiment has to be left overnight.)

Lipases

Make an emulsion of olive oil in a test tube and add a drop of litmus solution, or Universal Indicator and a drop of lipase enzyme. Leave it in a warm place for a day. Observe the reaction of the indicator as fatty acids form.

20.2 TYPES OF NUTRITION

20.2.1 Holozoic Types of Nutrition

This is the commonest type of nutrition found in the majority of animals. Food consisting mainly of macro-molecules of plant or animal origin is ingested, and broken down in the process of digestion to simpler molecules. These are then assimilated into the body, and undigested remains are egested as faeces.

EXPERIMENTS ON HUMAN BEINGS

Examine calorific values of common food materials and their vitamin content as obtained in diet tables. Design a balanced menu for two days, to give about 12,540 kilojoule (3,000 kilocalorie) food value, and a balance of vitamins. Dry samples of fuel foods give ;

Carbohydrates	17·6 kJ/g	(4·2 kcal/g)
Proteins	23·4 kJ/g	(5·6 kcal/g)
Fats	38·9 kJ/g	(9·3 kcal/g)

A balanced diet typically is made up of fuel foods giving ;

500 g carbohydrate, 100 g fat, and 100 g protein per day.

Keep complete records of food eaten over the course of a day or a week. Enter in a notebook the weight of each item of food throughout the day, and calculate the calorific value from diet tables. In the case of many foods containing a quantity of water, it will be necessary to know the original dry weight of food used, as this only constitutes the calorific value. For example, in weighing porridge, record only the amount of oats contained in a given portion of porridge. In making these records it will be useful to weigh previously the various plates on which food is placed, so that the amount of food can be recorded by weighing it on a plate of known weight. Make a chart of your results in the form of a histogram.

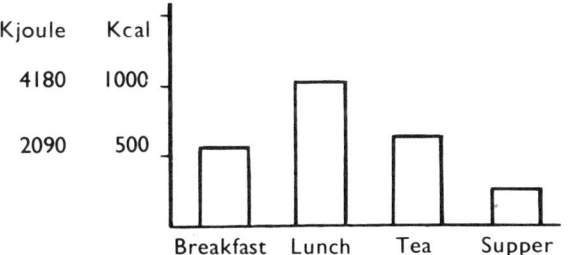

Calorific values from a calorimeter

A simple food calorimeter can be used for measuring the energy values of fuel foods. A weighed sample of food to be tested is placed in the crucible. Ignite the food quickly from a blow pipe and replace the calorimeter on the asbestos sheet. Take the temperature before the burning starts and note the maximum temperature when the burning is complete. 4·18 kilojoule is the quantity of heat required to raise the temperature of 1 kilogramme of water by 1°C. Thus by recording the weight of water in the calorimeter and the temperature rise in °C the calorific value of the food sample can be obtained (e.g. in an experiment 10 g of food when heated raised the temperature of water weighing 250 g by 40°C

$$\text{Food value} = \frac{250}{1000} \times 40 \times 4·18 = 41·8 \text{ kJ (10 kcal)}$$

Therefore the calorific value per gramme is

$$4·18 \text{ kJ/g or 1 kcal/g)}.$$

Compare the energy values of different food samples. Whereas the calorimeter itself is not completely accurate, the errors will be the same for each sample of food burned, and so the different values obtained for each sample can be compared.

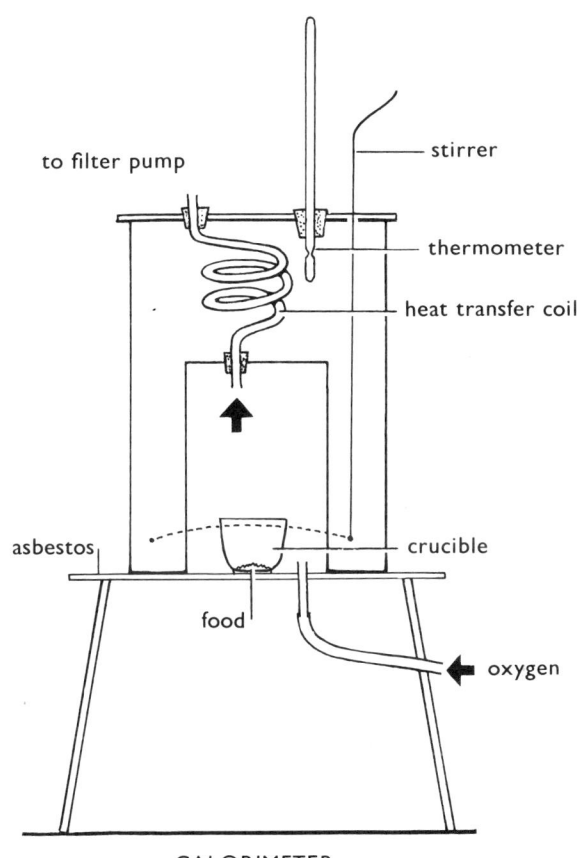

CALORIMETER

20.2.2 Holophytic Types of Nutrition

This is the commonest type of nutrition found in plants containing the green pigment chlorophyll. In such plants, food is made in the process of photosynthesis from the raw materials carbon dioxide and water, according to the equation :

$$6CO_2 + 12\,H_2O \xrightarrow{\text{light energy}} \underset{\text{glucose}}{C_6H_{12}O_6} + 6O_2 + 6H_2O$$

The carbohydrates formed in photosynthesis are normally converted into a food store of starch in the tissues of the plant. Most experiments on photosynthesis involve testing for starch as the end product of the process, and this can be done as follows.

Place a green leaf, such as a geranium or nettle leaf, in a beaker of industrial alcohol. Place the beaker on a water bath, and bring to the boil. Do not let any naked flame reach the alcohol. The boiling will remove the green pigments from the plant, the leaf now having a grey white appearance. Remove it from the beaker and place it in a dish of iodine solution. Those parts of the leaf containing starch will develop a purple-blue colour.

(a) To show that carbon dioxide is necessary for photosynthesis

Destarch a potted geranium plant by leaving it in a dark cupboard for 48 hours. Bring the plant into the light, and place one of its leaves in a conical flask containing a solution of concentrated sodium hydroxide, which keeps the atmosphere in the flask free from carbon dioxide. Make an air-tight connection by splitting a cork in half, with a single hole through the middle. Surround the petiole with this, and place the cork in the flask, making the joint air-tight with vaseline. Place the plant in light for 24 hours, and then carry out a starch test in a normal leaf, and in the leaf which has been kept in the carbon-dioxide-free atmosphere.

(Check that the plant was really destarched at the beginning of the experiment, by testing one of the leaves after the plant has been in the dark for 48 hours.)

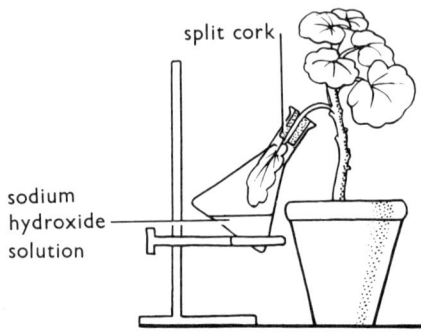

(b) Experiments on the intake of water into plants

ROOTS

Water enters a root by the root pressure developed by the physical process of osmosis. When water or a weak solution is separated from a strong solution by a semi-permeable membrane, water passes through the membrane into the strong solution, so diluting it. A true semi-permeable membrane allows water to pass through it, but not the molecules of the solute.

i. Experiments to show the physical process of osmosis

The Etherington osmometer. In this experiment the semi-permeable membrane is parchment paper.

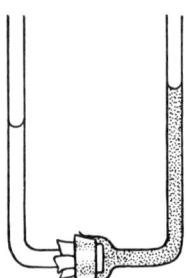

Visking tubing osmometer. Obtain some Visking (Cellulose) tubing, and construct an osmometer about 6 cm long as shown in the diagram. Fill the tube with 10% sucrose solution introduced through the top end from a syringe. Close the screw clip and suspend the tube full of sucrose solution from a balance arm. Dry its surface with filter paper and then record its weight. Immerse the tube in a beaker of water for about 15 minutes, then remove it from the beaker and again weigh the tube in air after it has been surface dried with filter paper. On completion of weighing, observe what happens when the tube is replaced in the beaker of water, and then open the screw clip. Carry out a second experiment, but this time immerse the Visking tubing containing 10% sucrose solution in a beaker containing 20% sucrose solution.

What conclusions can be drawn from these experiments?

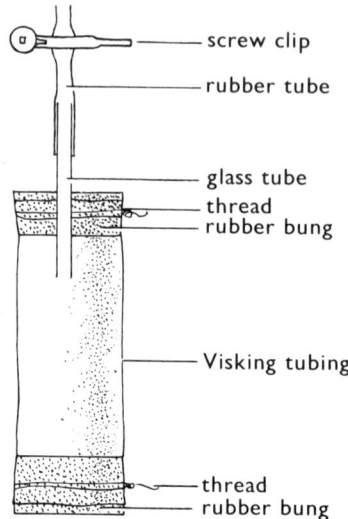

ii. Experiments to show osmosis in living cells

In cells, the cell membrane and cytoplasm which are alive act as semi-permeable membranes.

The potato osmometer. Place a teaspoonful of sugar in the centre of a potato in which a hollow has been made. Stand the potato in a dish of water, and observe the entry of the water through its tissues, to dilute the sugar inside the central hollow.

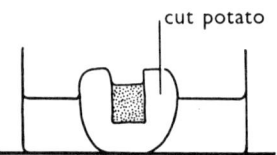

Cut out a cylinder of potato with a cork borer, and slice 20 discs from it, each about 2 mm thick. Weigh these, and record their total weight when fresh. Place the discs in distilled water for 1 hour, then remove them and dry their surfaces with blotting paper. Reweigh, and record the results. Repeat the experiment with 20 discs, which are placed in concentrated salt solution instead of distilled water, and compare the results of the two experiments.

Cut a large potato into strips about 7 cm long, and about 3 mm in cross-section. Place the strips in petri dishes over a piece of graph paper, dish 1 containing distilled water, dish 2 weak salt solution, dish 3 strong salt solution. Record the length of each strip at the beginning of the experiment, and after it has been left in the liquid for 30 minutes. Compare the results from each petri dish.

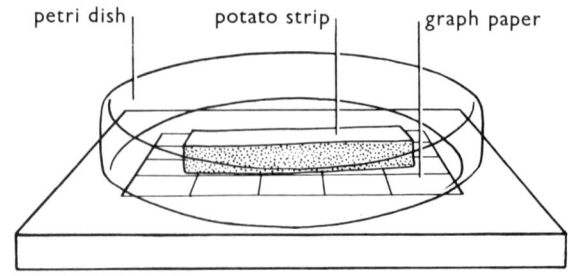

Cut the base of a dandelion stalk lengthwise with two cuts at right angles made with a razor, for a distance of about 5 cm. Place it in distilled water for 5 minutes, and then in strong salt solution for 5 minutes, and compare the behaviour of the stalk in each liquid.

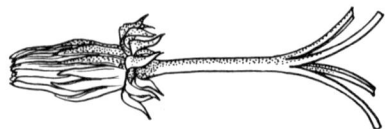

Many plant cells contain vacuoles which are easy to see when examined under the microscope, because they contain coloured solutions. Use the following specimens to investigate osmosis in plant cells.

Tradescantia stamen hairs (see 17.2.9) Take two microscope slides, on one place a few drops of distilled water, and on the other a few drops of 10% sodium chloride solution. Using forceps pick some hair cells from the stamen of a *Tradescantia* flower. Mount a few in distilled water on the one slide, and a few more in the 10% sodium chloride solution on the other. Observe what happens to the vacuoles in the cells on each of the slides over a period of a few minutes.

Beetroot cells Cut a strip of tissue from a fresh beetroot about 3 cm long and 3 mm cross section. Using a razor blade, cut some thin transverse sections. Place one is distilled water on a slide, and another in 10% sodium chloride solution on a separate slide. Observe what happens to the vacuoles over a period of a few minutes.

Geranium stigma hairs Repeat the experiments using hairs from the stigmae of Geranium flowers.

Record the behaviour of the cell vacuoles during the experiments, and suggest an explanation for what is observed.

iii. Experiments to show the entry of water into roots

Water entering roots by osmosis does so against the resistance of the plant tissues, and the effective height which the water reaches in the stem is an indication of the root pressure.

Cut a geranium plant off cleanly close to the soil in a pot, and immediately fix a capillary tube to it by means of a rubber connection. Observe the height water rises up the capillary tube by root pressure, until the roots die.

Thoroughly wash the roots of some cress seedlings. Place the plants so that the roots are immersed in a solution of red ink, and leave them in this for 24 hours. Then mount the whole of a seedling on a microscope slide, and observe which tissues in the root and stem have been stained by the coloured solution. (See 17.2.9)

STEMS

Experiment to show the physical process of capillarity

Study the way in which water rises in small tubes by placing three capillary tubes of different diameters in a beaker containing red ink.

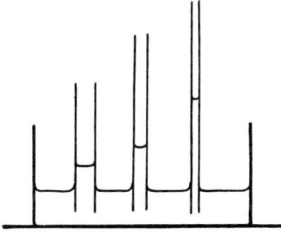

Place leafy shoots of willow or lilac in water. With a razor blade make a cut in the form of a ring round the stem so that it passes several mm into the wood. Observe the effect on the leaves.

LEAVES

Water escapes from the surface of leaves by a process called transpiration. This process differs from the physical process of evaporation in so far as the laws of evaporation refer to dead surfaces, but the cells of the plant leaf surface are alive, and so influence the rate of water loss.

Experiment to show the rate of evaporation

The rate of evaporation depends on temperature, air movement and humidity. Measure the rate of rise of the mercury in various conditions of temperature and air movement. It is important to make a firm connection between the porous pit and the cork. This can be achieved by vaselining the rim of the cork, and binding the porous pot with string, so that it is held tightly against the cork.

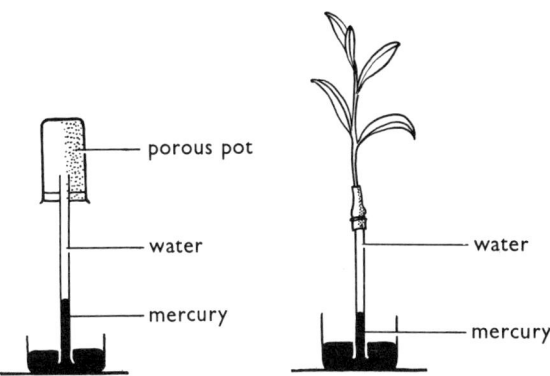

Experiments to show that water is lost from the surface of leaves

(i) Fix a piece of blue cobalt chloride paper by means of a paper-clip to the upper and lower surfaces of various leaves. Investigate which surfaces emit water.

(ii) Place a potted plant under a bell jar, covering the surface of the soil with a piece of polythene to prevent evaporation from the soil surface. Observe the condensation of water from the leaves under the bell jar. Place a similar plant from which the leaves have been removed under a similar bell jar, and compare the result with that of the leafy plant.

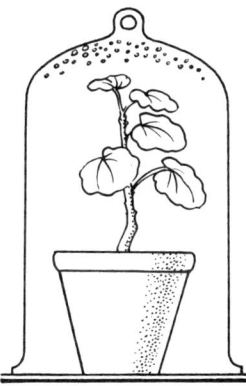

(iii) To show the presence of stomata in leaves. Place a leaf through a split cork, and insert it in a thistle funnel containing water. Invert this and attach it to a filter pump. Bubbles can be observed rising from the petiole, indicating that air is entering the stomata on the surface of the leaf.

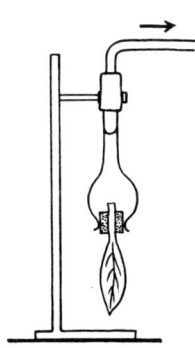

(iv) To measure the rate at which leaves lose water in air. Make a simple counterpoise balance (this can be improvised from a drinking straw) and suspend the leaf from one end, adjusting the balance so that it is horizontal. Place a 10 mg weight exactly over the leaf which will cause the balance arm to tilt. Measure the time that it takes for the balance arm to return to its horizontal position, and so record the time it takes for the leaf to lose 10 mg of water.

Then place the leaf on a piece of graph paper and trace its outline on to it. Estimate the total area of the two surfaces of the leaf and so obtain a value for the loss of water in mm²/hour.

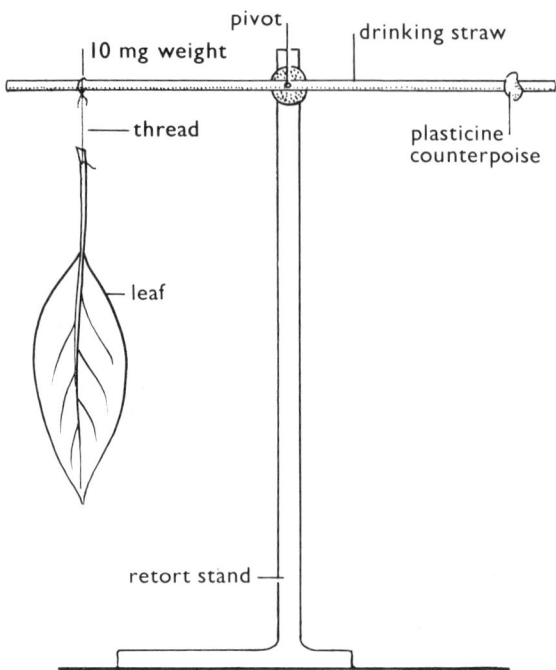

(v) Investigate the distribution of stomata in a leaf. On completion of experiment (iv) above cover the upper and lower surfaces of the leaf with a thin layer of nail varnish, and allow it to dry. Peel the nail varnish off each surface on to two separate slides and examine under the high power of the microscope. Calculate the area of the field of view in mm² and estimate the number of stomata which can be seen. Use this information to estimate the number of stomata per mm² on the upper and lower surfaces of the leaf. So from the results obtained in experiment (iv) above, estimate the weight of water lost from a stoma in an hour.

Experiments on the passage of water through shoots

(i) To measure the rate of transpiration from the leaves of a shoot, place a small shoot with about 6 leaves, freshly cut under water, into a specimen tube of water, through a loose-fitting cork. Weigh the whole on a balance, and then weigh again half an hour later. The loss in weight is equivalent to the amount of water transpired from the leaves.

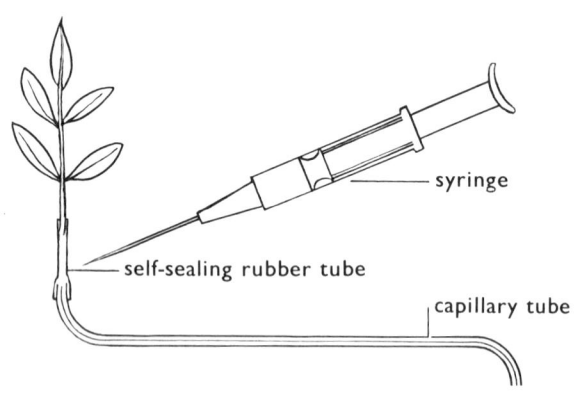

A SIMPLE POTOMETER

(ii) To measure the rate of absorption of water by a cut shoot, cut a shoot under water in a sink and place it in a potometer. A simple instrument can be made from a piece of capillary tube, to which a

small plant shoot can be attached by means of a piece of self-sealing rubber tubing. Mount the whole apparatus under water to exclude air bubbles, and ensure that the connection of the plant with the apparatus is airtight. Then allow a bubble of air to enter the capillary tube at the end opposite to the shoot, and measure the time it takes to move along a measured distance of about half the length of the tube. Next, insert the needle of a small syringe full of water into the rubber tubing, and force sufficent water through the system to drive the air bubble back to its starting point. Measure on the scale on the syringe body the amount of water needed to do this, and so obtain the volume of water absorbed by the plant shoot during the period of the experiment. Use this apparatus to measure the rate at which the shoot absorbs water in various conditions:

(a) a low and high temperature;
(b) calm, and moving air;
(c) dry, and humid air;
(d) in light, and shade;
(e) with the upper surface, or lower surface of the leaves vaselined.

Experiment to show guttation

In some plants, if the rate of osmosis in the roots exceeds the rate of transpiration from the leaves, water is forced out of the veins on to the tips of the leaves.

Germinate barley grains in specimen tubes of damp sand. When the leaves have developed, place the plant under a bell jar whose atmosphere is saturated with water. Observe drops of water forced out of the plant on to the leaf tips.

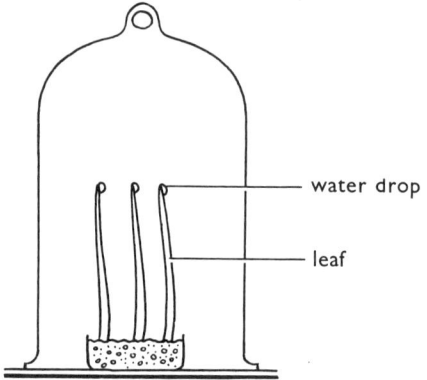

Experiments to show that light is necessary for photosynthesis

Destarch the leaves of several potted plants, such as geraniums, by keeping them in a dark cupboard for 48 hours.

(i) Bring four destarched plants out into the light, one for $\frac{1}{2}$ hr, one for 1 hr, one for $1\frac{1}{2}$ hr and one for 2 hr. At the end of each period remove a leaf, and test for starch using the iodine test. Determine which time of exposure is necessary before starch can be detected in the leaves.

(ii) Cover the leaves of a destarched plant with aluminium foil, in which various designs have been cut. Expose the leaf to light for several hours, then remove and test for starch. Observe which areas of the leaf have produced starch.

(iii) Destarch *Spirogyra* filaments by keeping them in the dark for several hours. Mount them on a slide, and expose to light for several hours. Place the filaments in cold industrial alcohol to kill them. Stain in iodine and examine under the microscope and observe the distribution of starch grains.

(iv) To show that different colours in the light spectrum influence photosynthesis. Place a destarched geranium plant in a dark box, and focus a spectrum through a slit in it on to a leaf. Expose to the spectrum for several hours, and at the end of this time test for starch formed in the leaf. Observe which part of the spectrum has induced the formation of starch.

(v) Place destarched plants in darkened boxes, and cover the top of each one with a different coloured filter. Investigate the colours which promote photosynthesis, by testing the leaves of the plants for starch formation.

(iv) Focus a spectrum on to a microscope slide containing destarched *Spirogyra*. At the end of several hours test for starch as in (iii) above.

To show that chlorophyll is necessary for photosynthesis

(i) Destarch the leaves of a variegated plant such as geranium or *Coleus*. Expose the leaves to light for several hours, and then test for starch. Observe in which part of the leaves starch is formed. Before carrying out the test for starch, map out the contours of the colours on the leaf by placing a piece of tracing paper on a sheet of glass over the leaf.

(ii) To show that chlorophyll itself is only formed in the presence of light, germinate two lots of cress seeds. Grow one lot in the light and the other in darkness. Observe that those raised in darkness remain colourless and are etiolated.

(iii) Make a solution of chlorophyll, by grinding fresh nettle leaves to a pulp in a mortar using a pestle. Add acetone to the pulp and then filter the solution through a Büchner filter funnel, using a filter pump. (An ordinary filter process can be used, but this will be slower.) Place the solution in a test tube and note that it is dichrooic, i.e. it looks green through the test tube, but red when light is reflected at the surface of the solution.

(iv) Place some spots of the solution on a dry filter paper, and note the rings of colour caused by the different pigments as the solution evaporates.

(v) Place a small quantity of solution in a watch glass, and stand a piece of blackboard chalk in it. Notice the pigments separating out at different levels, as the solution creeps up the stick of chalk by capillarity.

(vi) Set up a spectroscope to produce a complete spectrum, and place a tube of the solution in its path. Observe which colours in the spectrum are absorbed by the chlorophyll solution.

(vii) Grind up some mint or nettle leaves in acetone in a mortar. Filter through a Büchner funnel and collect the filtrate, which will contain a solution of the chlorophyll pigments. Cut a thin strip of Whatman No. 1 filter paper and suspend it from the lid of a gas jar by securing it with adhesive tape. The strip should be about 0·5 cm from the bottom of the jar when the lid is in position. Make a pencil line across the filter paper about 3 cm from the bottom end, and on to this line load a spot of the chlorophyll in acetone solution. Then place some petroleum ether in the gas jar so that it is about 1 cm deep, and hang the filter paper so that it is dipping into it. Ensure that the lid of the gas jar makes an airtight connection, using vaseline if necessary. The petroleum ether will creep up the filter paper carrying the chlorophyll pigments from the loaded spot with it. When it has almost reached the top, remove the filter paper and allow it to dry in the atmosphere. Observe the separation of the chlorophyll pigments at different levels.

As the chemicals are very inflammable, do not have any naked flames in use during these experiments.

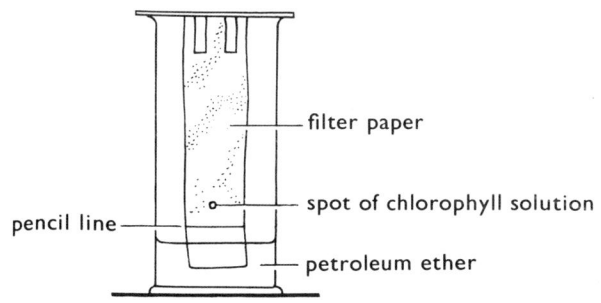

filter paper
spot of chlorophyll solution
pencil line
petroleum ether

Experiments to show that oxygen is produced in photosynthesis

(i) Place a geranium plant inside a bell jar, together with a small candle. Light the candle so that it uses up all the oxygen. Ensure that the bell jar is tightly sealed with vaseline to a glass plate. Expose the plant to light for a day, then introduce a lighted taper to test for the presence

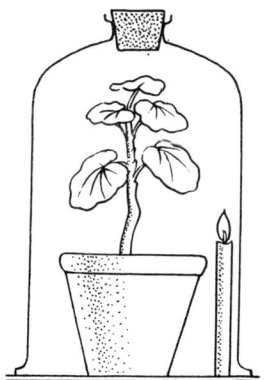

of any oxygen formed by the plant. Carry out a control experiment by placing a similar piece of apparatus in the dark.

(ii) Place a water plant, such as Canadian Pondweed (*Elodea* spp.) under a funnel in a beaker, setting up the experiment under water in a

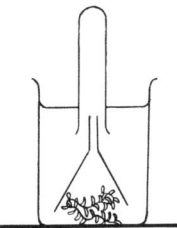

basin, to exclude air bubbles at the outset. Place the apparatus on a bench and expose it to continuous light from an electric bulb for 48 hours. Test the gas which collects in the test tube.

(iii) Insert the stem of a water plant through a hole in a small cork, and fit this into an ignition tube full of water. Invert the whole apparatus inside a large beaker of water, fixing the ignition tube to a retort stand by means of a clamp. Observe the amount of oxygen collected in the ignition tube at a given time, when the plant is exposed to:

White light.
Light of different colours.
When the beaker contains distilled water.
When the beaker contains a weak solution of sodium bicarbonate.
White light, with the water at different temperatures.

Alternatively, count the number of bubbles emerging from the cut end of the stem in a given time.

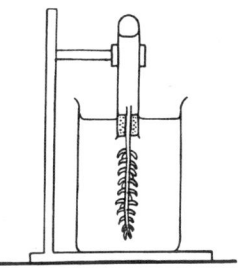

Capillary tube method for measuring small gas volumes

In the experiments (ii) and (iii) above, the presence of carbon dioxide and oxygen can be tested using a piece of capillary tube, with a brass screw attached suitably greased to make it airtight. (See diagram.) The principle of the tests depends on the fact that carbon dioxide gas is absorbed by strong potassium hydroxide solution, and oxygen is absorbed by alkaline pyrogallol.

To take a gas sample, turn the screw fully inwards, and then draw a column of water about 5 cm long into the tube, by turning the screw

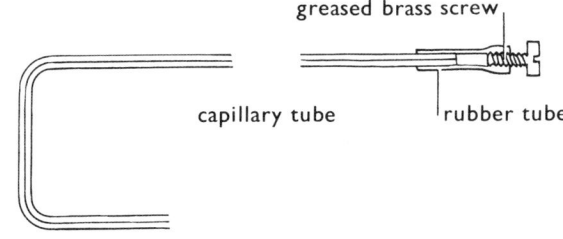

greased brass screw
capillary tube
rubber tube

GAS SAMPLING APPARATUS

slowly outwards. Then insert the end of the apparatus into the test tube containing the gas to be sampled, and draw up about 10 cm length of gas. Seal this in by placing the end of the apparatus under water again, and draw in a short column of water. As the volume of gas inside the apparatus will be affected by temperature changes, it is essential to keep the temperature constant by placing the whole apparatus in water near room temperature in a sink or in a water bath. Measure the length of the gas sample after the apparatus has been at a constant temperature for about 5 minutes.

Next place the open end of the capillary tube in a strong solution of potassium hydroxide, and carefully turn the screw until the gas sample touches the potassium hydroxide, being very careful not to lose any gas. Unscrew and draw a little potassium hydroxide solution into the capillary tube, and move it backwards and forwards to make a good contact with the gas sample. Leave the apparatus to stand in the water bath for about 5 minutes and then measure the length of the gas bubble. A decrease in length would indicate the presence of carbon dioxide in the original sample.

Next, carefully expel the potassium hydroxide solution from the apparatus under the surface of some alkaline pyrogallol, and replace it with a short length of the pyrogallol solution. Gently mix it with the gas by moving the bubble backwards and forwards along the capillary tube. Again leave the apparatus in a water bath for about 5 minutes and then measure the length of the gas bubble. A further shortening in length indicates the presence of oxygen in the original sample.

Intake of minerals by plants

In order to synthesize organic food materials, traces of minerals are necessary in addition to carbon dioxide and water. Germinate cress seedlings, and when the roots are approximately 1 cm long place the roots through holes in circular cards, cut to fit over the top of 250 cm³ gas jars. Set up 9 jars with several seedlings on a card on top of each, and fill the gas jars with various solutions listed below, so that the roots enter the liquid. It is important that the gas jars should be clean, before putting the solutions in them, and the jars should be covered with black paper during the experiment to protect some of the chemicals from deterioration, and to prevent the growth of algae.

Put liquids in the jars as follows:

1. Distilled water.
2. Culture solutions made up from the following quantities dissolved in 250 cm³ distilled water.
   Potassium nitrate, KNO_3,—0·2 g
   Calcium sulphate, $CaSO_4 . 2H_2O$,—0·07 g
   Calcium hydrogen phosphate, $CaHPO_4 . 2H_2O$,—0·07 g
   Magnesium sulphate, $MgSO_4 . 7H_2O$,—0·07 g
   Sodium chloride, $NaCl$,—0·02 g
   A few drops of ferric chloride, $FeCl_3$, a trace of manganous chloride, $MnCl_2$, and boric acid, H_3BO_3.
3. A culture solution lacking potassium. (Replace KNO_3 with $NaNO_3$,—0·02 g)
4. A culture solution lacking calcium. (Replace $CaSO_4 . 2H_2O$ by K_2SO_4,—0·05 g; and $CaHPO_4 . 2H_2O$ by $Na_2HPO_4 . 12H_2O$,—0·02 g)
5. A culture solution lacking magnesium. (Replace $MgSO_4 . 7H_2O$ by K_2SO_4,—0·05 g)
6. A culture solution lacking iron. Omit ferric chloride, $FeCl_3$.
7. A culture solution lacking nitrogen. (Replace KNO_3 by KCl,—0·15 g.)
8. A culture solution lacking phosphorus. (Replace $CaHPO_4 . 2H_2O$ by $Ca(NO_3)_2$,—0·04 g.)
9. A solution lacking sulphur. (Replace $CaSO_4 . 2H_2O$ by $CaCl_2$,—0·04 g; and $MgSO_4 . 7H_2O$ by $MgCl_2$,—0·05 g.)

Tablets can be obtained from laboratory suppliers containing the necessary chemicals for these experiments, and these have only to be dissolved in an appropriate volume of distilled water to produce solutions of the required strength.

Experiments on soil

Obtain samples of soils from various habitats, and at least two depths. Use either a spade or a soil augur to obtain the samples.

i. Mineral content

Shake up a sample of soil in a boiling tube and allow it to settle. Estimate the percentage of the different particles, by placing a ruler against the side of the tube, and measuring the size of each layer.

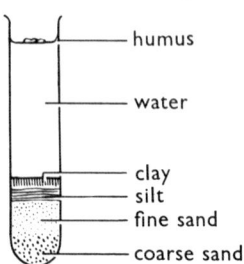

— humus

— water

— clay
— silt
— fine sand

— coarse sand

Using a pipette with a fine point draw in a small sample of soil from the coarse sand layer, and mount a few grains on a microscope slide. Record and measure their size. Repeat the experiment by studying samples from each of the other layers and so compare the microscope structure and size of the main mineral ingredients of the soil sample.

ii. Water content

Weigh a clean evaporating dish, and place a sample of fresh soil in it. Weigh again in order to obtain the weight of the soil sample. Dry the soil by placing it in an oven at about 102°C, or by warming it gently on a sand tray. Weigh the sample again after heating has continued for 30 minutes. Estimate the percentage of water which was present in the sample of soil.

Estimate the water content of saturated soils, watering the soils in their natural habitat, before removing them to the laboratory for examination.

iii. Water-holding properties

Put samples of clay, loam and sand into a muslin bag each and saturate with water. Allow the surplus to drain away, and then place each sample in a weighed evaporating basin. Leave the samples in a warm room, and weigh at intervals. Determine which samples retain water most effectively.

iv. Air content

Take a 7 lb jam tin, from which the lid has been cleanly removed, and pour water into it from a measuring cylinder, to determine its volume. Lower the tin into a tall aquarium tank, and mark the water level with a label. Remove the jam tin full of water, and then tip the water out of it. Punch several holes in the bottom of the tin with a small nail. Take the tin, and carefully push it into a sample of soil, taking care not to compress the soil. Dig the whole out with a spade, and gently level off the soil in the tin. Place the tin back into the aquarium tank, and irritate it so that all the air bubbles escape from the soil. Mark the new level of water with a label on the side of the tank. Top up to the first level with water from a measuring cylinder, the volume of water added is equivalent to the volume of the air in the soil sample. Record the percentage of air in the soil sample.

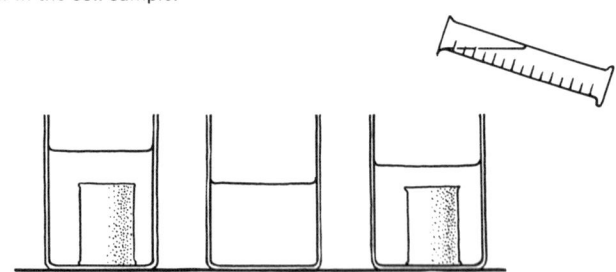

v. Humus

Take a sample of soil, and weigh it in an evaporating dish after it has been dried in an oven at 102°C for about 30 minutes. (The sample left from section (ii) above can be used for this experiment.) Then heat the dish strongly over a bunsen burner to burn all the humus content. Weigh again when cool, and so estimate the percentage of humus present in the fresh soil sample.

vi. Reaction

Place a small amount of soil in an evaporating basin, and add a few drops of Universal Soil Indicator. Estimate from the colour which develops whether the sample is acid, neutral or alkaline.

vii. Minerals

Add distilled water to a sample of soil, shake and filter. Test for the metals and acid radicals present in the soil, as described in 20.1.2.

viii. Carbonate content

Place a sample of soil in the bottom of a conical flask. Fit a cork with a delivery tube, and suspend an ignition tube from a piece of thread inside

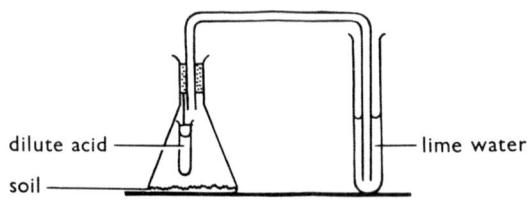

dilute acid —

soil —

— lime water

the flask, containing dilute hydrochloric acid. Shake the acid on to the soil sample, and test whether carbon dioxide is produced by allowing any gases formed to pass through a tube containing lime water.

ix. *Physical properties*

(a) Permeability to water Set up three samples of soil (sand, silt and clay) in filter funnels, supported on glass wool. Pour an equal volume of water through each one, and measure the time for it to pass through in each case.

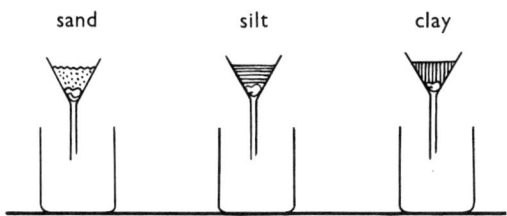

(b) Permeability to air Set up three samples of similar soils (sand, silt and clay) in filter funnels, attached by rubber tubing to the top of a burette, previously filled with water. Open the burette taps in turn, and measure the time it takes the water to leave, so giving a measure of the permeability of each soil sample to air.

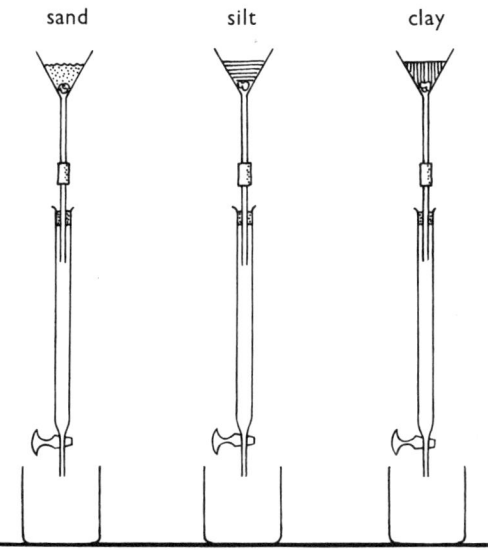

(c) Capillarity Fill three wide tubes with sand, silt and clay, which have been oven dried. Ensure that the specimens are well packed into the tubes. Cover the bottom of each tube with a piece of muslin to prevent the soil from falling out, and place the tubes in dishes of water. Observe the rate at which water ascends each tube by capillarity.

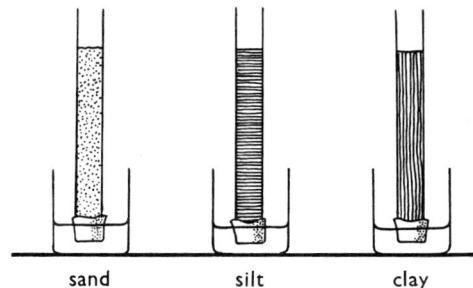

x. *Soil organisms*

(a) Place a fresh soil sample in a slightly inclined measuring cylinder, and introduce a slow flow of water into the bottom through a glass tube. Small organisms will be washed out of the soil, and can be collected for examination in a fine silk sieve.

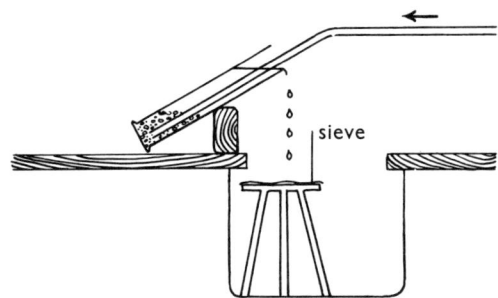

An alternative method is to place a sample of fresh soil in a Tullgren funnel and leave it for up to 24 hours. A suitable apparatus can be made from a funnel of smooth shiny paper fixed to a ring clamp on a retort stand. A sieve of fine mesh is fixed inside the funnel and the fresh sample of soil is placed on the sieve, leaving a clear space all round it. Fix a 40 watt electric light about 20 cm above the sample and leave it on for the period of the experiment. Any animals present in the soil will move away from the source of heat and slip down the side of the funnel into a collecting beaker. A weak solution of formaldehyde should be placed in the beaker to preserve the animals until they can be sorted out for identification and study under the microscope.

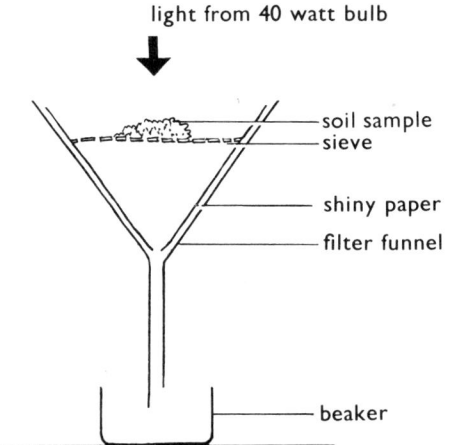

SIMPLE TULLGREN FUNNEL

(b) Place three layers of distinct kinds of soil in a glass aquarium, and cover with dead leaves. Introduce several earthworms into the soil, and observe how they mix up the layers over a period of time.

(c) Place one sample of fresh soil in a small muslin bag, and suspend it inside a corked conical flask containing lime water. Observe the change in colour as it absorbs carbon dioxide given out by the soil organisms. Set up a control experiment in which the soil sample has previously been sterilized by heating to boiling point.

xi. *Soil profiles*

Examine the profile of natural soils by digging a pit with vertical sides. Carry out soil tests on samples from different depths.

xii. *Soil temperature*

Measure the soil temperature, by inserting the bulb of a thermometer into the side of a freshly dug pit at different depths.

xiii. *The effect of lime*

Place a sample of clay soil in a measuring cylinder, and shake it up in water. Add a few gm of calcium oxide, and observe the effect on the clay suspension. Set up a control experiment from which lime is omitted.

20.2.3 Saprophytic Types of Nutrition

Saprophyric organisms, mostly Fungi and Bacteria, obtain energy by breaking down dead organic materials.

(i) Allow different types of food substances, such as bread, fruit, jam, etc., to develop moulds by placing them under a bell jar which is kept humid and warm. Examine the organisms which develop on these foods.

(ii) Examine the growth of yeast in a sugar solution. (See 21.2.)

(iii) Allow dead flies to develop moulds, by leaving them in water. Species of the fungus *Saprolegnia* (see 14.4.2) may develop. Examine the reproductive organs in which motile zoospores can often be seen.

20.2.4 Parasitic Types of Nutrition

Parasitic organisms obtain their food supply from another living organism—the host.

Grow cress seedlings in dishes under a bell jar, and allow the atmosphere to become warm and humid. Examine the fungi which develop on the seedlings and kill them. (Often species of *Pythium*.)

20.2.5 Symbiotic Types of Nutrition

An association of two different organisms whose feeding habits help each other.

Examine the nodules on the roots of lupins which contain bacteria able to convert nitrogen into nitrates, and are beneficial to the lupin plant. Squash a small piece of nodule on a slide and examine it under the high power of the microscope. (See page 82.)

Examine Lichens which may be found growing on the trunks of trees, or on exposed rocks. Many of these consist of an Ascomycote fungus associated with a simple spherical type of green alga similar to Pleurococcus (Page 57). Squash a small piece of Lichen on a slide and examine it for the presence of the fungal hyphae and the green algae.

Examine the green hydra *Chlorohydra viridissima* (Page 13) which associates with the cells of the green alga *Zoochlorella* contained in the hydra's endoderm cells.

20.2.6 Food Storage Organs

Carry out tests for the types of food stored in food storage organs using the tests described in 20.1.2. Test for food substances stored in: potato, bulb, corm, carrot, grass rhizome. Fruits of sunflower, wheat and barley, seeds of castor oil, pea and bean.

---

21 Experiments on Respiration

Respiration is a process by which living organisms obtain energy through the breakdown of molecules of organic materials. Most organisms are aerobic, i.e. they require oxygen to facilitate the process, but a few are able to break down their food molecules without oxygen, and are said to be anaerobic. In both processes, the actions are started and catalysed by respiratory enzymes.

Aerobic respiration

$$\frac{\text{Food}}{\text{substrate}} + \text{Oxygen} \rightarrow \frac{\text{Carbon}}{\text{dioxide}} + \text{Water} + \text{Energy}.$$

$$\underset{\text{glucose}}{C_6H_{12}O_6} + 6O_2 \rightarrow 6CO_2 + 6H_2O + \underset{\text{(674 kcal)}}{2830 \text{ kjoule}}$$

Anaerobic respiration

$$\frac{\text{Food}}{\text{substrate}} \rightarrow \frac{\text{Carbon}}{\text{dioxide}} + \text{Ethyl alcohol} + \text{Energy}.$$

$$\underset{\text{glucose}}{C_6H_{12}O_6} \rightarrow 2CO_2 + 2C_2H_5OH + \underset{\text{(26 kcal)}}{109 \text{ kjoule}}$$

The energy released from the breakdown of glucose is not directly used by living organisms, but most of it is stored inside the cells in a complex compound called adenosine triphosphate (ATP for short). This is formed by the addition of phosphate to adenosine diphosphate (ADP), a similar compound, using the energy obtained from the breakdown of glucose.

$$\text{ADP} + \text{Phosphate} + \text{Energy} \rightarrow \text{ATP}$$

When the organism requires energy it can be released quickly from the ATP, which is then converted back to ADP with the formation of phosphate.

$$\text{ATP} \rightarrow \text{Energy} + \text{Phosphate} + \text{ADP}$$

So the process is a cycle which can be represented as follows:

```
   ENERGY          ADP         ENERGY
from glucose      ( P )      for life processes
 breakdown          ATP      e.g. movement, growth
                             irritability.
```

21.1 AEROBIC RESPIRATION

ANIMALS

i. To show that oxygen is absorbed

Place an active animal such as a cockroach or earthworm inside a conical flask. Fit a U-shaped piece of capillary tube to it through a cork, with its free end dipping into water. Suspend a tube of sodium hydroxide solution inside the flask. As carbon dioxide is evolved it is absorbed by the sodium hydroxide, so water rises up the capillary tube as the oxygen is absorbed by the animals. In order to avoid changes of pressure inside the flask caused by slight variations of temperature outside, this experiment must be carried out in a water bath. Carry out similar experiments at different temperatures, and record the rate of oxygen absorption by means of a graph.

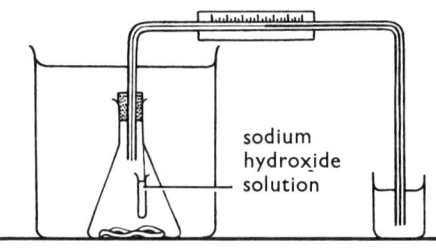

sodium hydroxide solution

ii. To show that carbon dioxide is evolved

(a) Suspend a tube of lime water in a conical flask, containing an active animal. Close the flask with a cork, with a short length of capillary tube passing through it. Observe milkiness in the lime water as carbon dioxide is evolved. Set up a control experiment in a flask containing lime water only.

(b) Place a few small fish in a bowl of fresh water and leave for 24 hr, together with a control dish without any animal in it. At the end of 24 hr, place some fresh lime water into each of two test tubes. Into one place some drops of water removed from the bowl of fish, and into the other place drops of water obtained from the control.

(c) To show that there is more carbon dioxide in exhaled air than in inhaled air set up the apparatus shown in the following diagram, and breathe through it steadily.

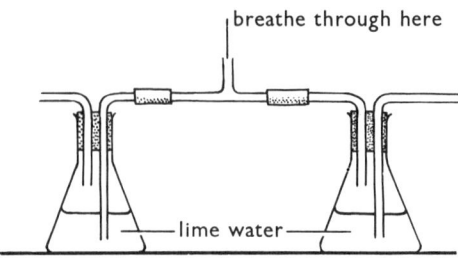

breathe through here

lime water

iii. To show that water is evolved

(a) Breathe on to a cold mirror, and observe moisture condensing.

(b) Breathe on to a piece of dry cobalt chloride paper and observe the change in colour as it becomes damp.

iv. To show that heat is evolved

Place several earthworms on damp cotton-wool inside a thermos flask and record the temperature with a sensitive thermometer over the course of a few hours. Compare this with a similar flask in which there are no worms.

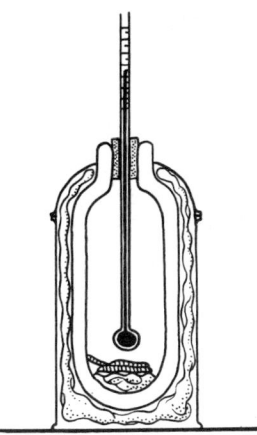

v. To show the evolution of light

Slit a fresh herring in half, and leave on a dish under a bell jar in a warm, dark cupboard. Examine in a dark room after a few days, when light may be seen emitted by bacteria respiring on the tissues of the fish.

vi. To show the action of ATP

Obtain a piece of fresh lean meat, and with a pair of mounted needles separate out a fine strand of muscle fibres about 3 cm long. Lay this carefully on a clean microscope slide and measure the length of the muscle tissue accurately. Add a few drops of ATP solution to the meat specimen, and measure its length again after about 1 minute.

Compare the action of ATP with that of distilled water and glucose solution by carrying out two similar experiments. In the first, place a few drops of distilled water on a fresh meat strand, and in the second a few drops of glucose solution. In each case investigate whether these fluids cause any change in the length of the meat fibre.

vii. Respiration mechanisms

(*a*) Pond snail (*Limnaea* spp., see 11.2). Observe an animal taking air at the surface of water.

(*b*) Crayfish (*Astacus fluviatilis*, see 10.1). Observe the breathing current by placing a spot of ink from a pipette at the posterior lower surface of a gill cover.

(*c*) Mussel (*Mytilus edulis*, see 11.1 and 18.2). Observe the current of water caused by the gills, by placing a piece of gill in sea water in a petri dish. Place specks of dust on the gill, and observe the movement.

(*d*) Gnat larvae (*Culex* spp., see 10.2.5). Observe animals breathing at the surface of water.

(*e*) Minnow (*Phoxinus laevis*, see 13.1.2). Observe minnows gulping air at the surface of water.

(*f*) Frog (*Rana temporaria*, see 13.2). Observe fully developed tadpoles gulping air at the surface of water. Observe an adult frog breathing, and notice the different movements when the animal is breathing through the lining of the mouth cavity, or through its lungs. (See page 35.)

(*g*) Investigate the capacity of your lungs. A normal adult breathing quietly moves about 500 cm^3 of air in and out of the lungs and this is called the *tidal air*. If he then inhales as deeply as possible, a further 1,500 cm^3 of *complemental air* is breathed in. It is also possible when breathing quietly to exhale more than the usual amount of air and this is called the *supplemental air*. The three components when added together constitute the *vital capacity* of the lungs, which represents the maximum volume of air which can be dealt with by breathing movements. It is not possible to exhale completely, and the small amount of air remaining in the lungs is called the *residual volume*. These relationships can be shown as follows:

```
Maximum inspiration ──────────────────5000 cm³
                   ┌ ─────────────────
                   │  COMPLEMENTAL AIR
                   │ ──────────────────3500 cm³
VITAL              │  TIDAL AIR
CAPACITY          ┤ ──────────────────3000 cm³
                   │  SUPPLEMENTAL AIR
Maximum expiration └──────────────────1500 cm³
                      RESIDUAL VOLUME
                   ─────────────────── 0 cm³
```

Two types of apparatus can be used for these investigations. The simplest consists of a large calibrated bell jar inverted over water either in a sink, or in a large pneumatic trough. Fill it with water by connecting one outlet to a filter pump. Then close the connection to the filter pump, put the breathing tube in your mouth, release the clip and breathe into the bell jar so displacing a volume of water equal to the amount of air exhaled.

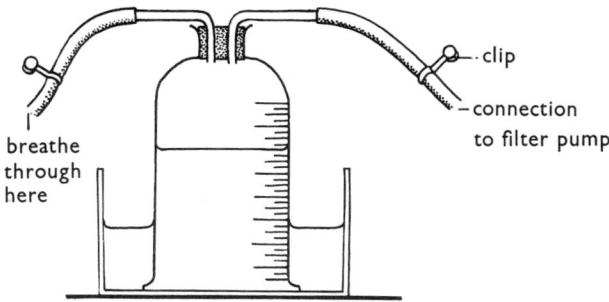

A more efficient apparatus can be made from two large tins, an outlet valve and a non-return valve, as shown in the diagram.

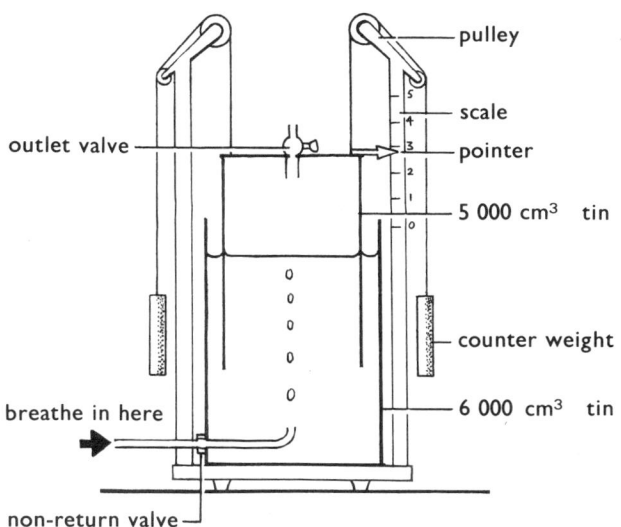

RESPIROMETER

(*h*) Remove the lungs and trachea from a dissected rat without damaging them. Place them in a bell jar, connecting the trachea to a glass tube passing through a cork. Ensure that the connections are airtight. Place a rubber sheet over the open end of the bell jar to represent the diaphragm. Move this artificial diaphragm upwards and downwards, and observe the action of the lungs.

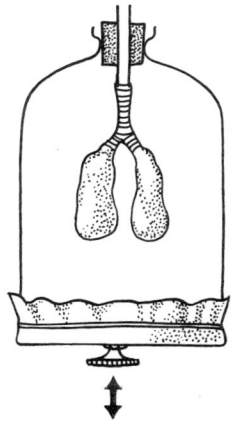

(*i*) There is evidence that cigarette smoking has a harmful effect on the lungs. In order to reveal some of the substances present in cigarette smoke, set up an apparatus which will draw the smoke through a condenser from a burning cigarette.

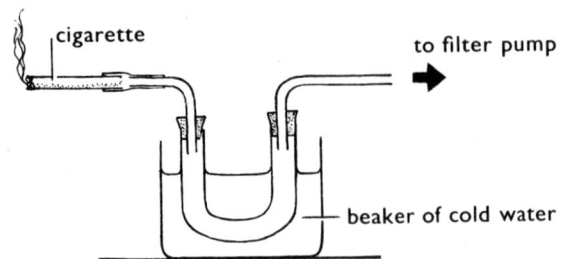

viii. Circulation mechanisms

(*a*) Earthworm (*Lumbricus terrestris*). Investigate pulsation in the dorsal blood vessel as described in 18.2.

(*b*) Water flea (*Daphnia pulex*). Mount a living specimen under a coverslip on a cavity slide, and trap it in a small amount of cotton wool so that it cannot move. Focus the heart under the microscope and watch the circulation of blood corpuscles in and around it. (See page 21.)

(*c*) Trout. Study the circulation in a trout alevin by mounting a specimen in water in a cavity slide. Examine the circulation in the region of the gills and tail under the microscope.

(*d*) Tadpole (See page 37). Mount a young tadpole in the external gill stage on a cavity slide in a drop of water. Examine the circulation in the gills and tail under the microscope.

ix. Mammalian blood

(*a*) Obtain a small drop of blood, by pricking the thumb just below the nail with a sterilized needle. Make a smear on a slide (see pages 6 and 51) and examine the blood cells in the sample under the high power.

(*b*) Using a blood count apparatus, measure the number of red corpuscles. Suck a small sample of blood into a pipette to the 0·5 mark and make up to the 10·1 mark by sucking up Ringer's Solution, so diluting the sample by 1/200. Shake thoroughly, then place a few drops on the calibrated slide of the apparatus. (Squares of 1/400 mm² and 1/10 mm deep.) Count the average number of corpuscles in several squares, and so estimate the number in a cubic mm of blood of the original sample. *N.B.* One square contains 1/4000 cm³ of blood, which has been diluted 200 times.

(*c*) Clotting. Place a small drop of blood on a slide, and leave it for 15 minutes. Then examine it for the formation of fibrils. Place another drop of blood on a slide to which a drop of 10% sodium citrate has been added. This prevents clotting.

(*d*) Place a small sample of blood in a test tube to which a little 10% sodium citrate has been added, and dilute it 20 times with distilled water. Bubble a stream of oxygen through it, and observe the colour change. Subsequently observe the colour change when carbon dioxide is bubbled through the sample.

(*e*) Test for blood. Place a drop of blood in a test tube and dilute it with distilled water. Add tincture of guiacum and hydrogen peroxide, and observe the colour which develops.

(*f*) Blood groups. It is possible to identify the human blood groups A, B, AB, and O by using prepared sera of two kinds: 'Anti-A' and 'Anti-B'. If a serum reacts with a blood sample the corpuscles come together in clumps which can be recognized under the microscope. If there is no reaction the corpuscles stay in a dispersed condition. Take a clean microscope slide and place it on a white tile. Label the left-hand edge A and the right-hand edge B. Place one spot of fresh blood at each end of the slide. Then mix a drop of serum 'Anti-A' with the left-hand spot using a clean match stick, and a drop of serum 'Anti-B' to the right-hand blood spot, using another clean match. After a few seconds, examine the two samples to see whether there is any clumping. Work out your blood group.

PLANTS

i. To show that oxygen is absorbed

Repeat the experiment described for animals in paragraph (i), page 112, using germinating peas placed on damp cotton-wool inside the conical flask.

ii. To show that carbon dioxide is evolved

(*a*) Repeat the experiment described for animals in paragraph (ii), page 112, using germinating peas placed on damp cotton-wool inside the conical flask.

(*b*) Set up a plant under a bell jar, placed on a piece of glass, and vaselined to make an air-tight connection. By means of a filter pump draw air through the apparatus, carbon dioxide being removed from it by means of a tube containing soda lime. Test that carbon dioxide is evolved due to the plant respiration, by passing the air exhaled from the plant through a tube of lime water. It is necessary to cover the bell jar with a black cloth, to prevent photosynthesis taking place and interfering with the experiment.

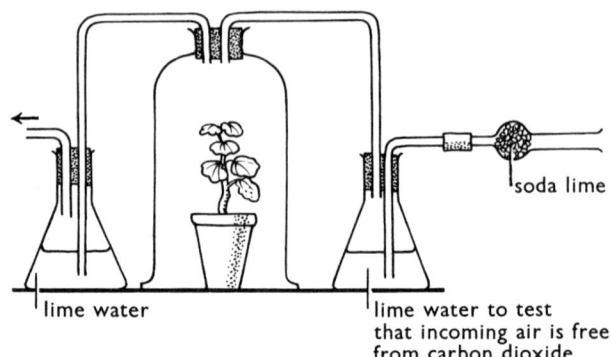

(*c*) The indicator phenolphthalein is very sensitive to the presence of weak acids. It is pink in alkaline solutions and colourless in acid. Place two germinating peas with roots just emerging in a test tube, and just cover them with 0·005% sodium hydroxide solution; add one drop of phenolphthalein. A faint pink colour will form but will quickly disappear when the peas have produced enough carbon dioxide to neutralize the weak solution of alkaline sodium hydroxide. The rate of carbon dioxide production can be measured at different temperatures by repeating this experiment by placing test-tubes set up as described above in water baths. Temperatures of 5°C, 15°C, 25°C and 35°C would be suitable. In each case measure the time taken for the indicator to decolorize.

iii. To show that water is evolved

Repeat the experiment as in (ii *b*) above, replacing the soda lime tube with calcium chloride contained in a similar tube, to dry the incoming air. Observe the formation of water condensing inside the bell jar. The soil should be covered with a polythene sheet.

iv. To show that heat is evolved

Place several germinating peas on damp cotton-wool inside a thermos flask, and record the temperature with a sensitive thermometer over the course of a few hours. Compare this with a control flask in which there are no seeds.

21.2 ANAEROBIC RESPIRATION

This process takes place in those parts of plants which are unable to obtain oxygen, and also in some fungi and bacteria.

i. Anaerobic respiration in germinating peas

Set up the apparatus as shown, placing germinating peas in the test tube, so that they float to the top of the mercury. After 2 days test the gas which they evolve into the test tube. Set up a control experiment, in which the peas have previously been killed by heating.

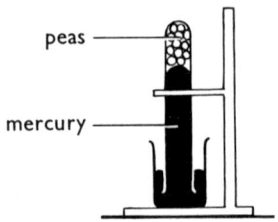

ii. The fermentation of yeast

Place a few cm³ of brewers' yeast in a solution of glucose contained in a conical flask. Connect this by means of a tube, dipping into a test tube of lime water. Leave in warm place (25°–30°C) for 2 days. At the end of this period observe milkiness in the lime water, and the smell of ethyl alcohol in the flask. If a portion of this is distilled, sufficient alcohol will be produced to burn in a watch-glass. Set up a control experiment, in which the yeast has previously been killed by heating.

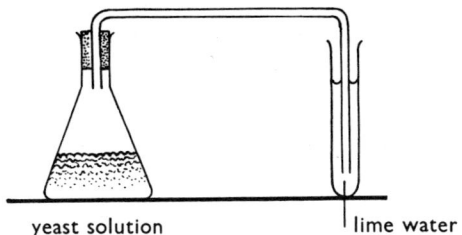

yeast solution   lime water

21.3 RESPIRATORY ENZYMES

i. Dehydrogenases

Place a little milk in a test tube together with a few drops of methylene blue. Leave in a warm place for several hours. The methylene blue is decolorized as it is reduced by bacterial respiration in the milk.

ii. Peroxidases

Leave a piece of cut potato exposed to the air. It goes brown due the activity of these enzymes in the presence of oxygen in the atmosphere.

Grind up a little fresh potato with a pestle and mortar, and add a few drops of tincture of guiacum. It turns blue owing to oxidation brought about by enzymes in the potato.

22 Experiments on Excretion

Excretion is the process in which the waste products of metabolism are separated from the tissues of an organism. In animals excretory products are expelled from the body into the environment, but in most members of the plant kingdom, the excretory products are retained within the tissues. In animals the main excretory products are ammonia, urea and uric acid; in plants a wider variety of substances is produced.

22.1 ANIMAL EXCRETION

i. Earthworm (*Lumbricus* spp.)

Examine nephridia removed from a freshly killed worm as described on page 19.

ii. Cockroach (*Periplaneta americana*)

Remove a few malpighian tubules from the dissection of a cockroach as described on pages 22 and 23, and examine them on a microscope slide, under low and high power lenses.

Place some cockroach droppings in a small hard glass tube and heat with soda lime. Test the gas evolved, by placing a piece of damp litmus paper near the opening of the tube.

iii. Man

Obtain some sweat by wearing a rubber finger stall for a short time. Pick up a specimen of sweat from inside the stall on a piece of platinum wire, and test for sodium using the flame test. Test for chlorine by placing a sample in a test tube containing silver nitrate.

22.2 PLANT EXCRETION

i. Fungi

Grow a culture of *Mucor* on a petri dish on agar jelly in which a little sugar has been dissolved. After several days test the reaction of the jelly for staling products, by removing a little of it and dropping into it a solution of neutral litmus in a test tube.

ii. Seaweeds

Boil a sample of *Fucus* spp. in a large beaker. Filter through a small mesh wire sieve. Observe the jelly produced when the extract cools.

iii. Coniferophyta

Obtain resin from the trunk of a pine tree and test its reaction to litmus. Make a solution by warming some resin in water. Test its antiseptic properties by placing a few drops on part of an agar jelly surface contained in a petri dish exposed to the air. Observe whether bacteria and fungi develop on the part of the agar containing the resin solution.

iv. Anthophyta

A wide variety of substances is obtained from members of this division, many of which form the basis of medicinal drugs.

(a) *Tannins*. These are present in tree bark, or tea leaves. Boil a small portion of the skin of a rat in strong tea, and observe the tanning effect.

Boil a little bark from an oak tree with ferrous sulphate, and so make a weak solution of ink. Write with this, and observe that it is several days before the writing blackens. (Blue-black ink contains an extra blue pigment.)

(b) *Oils*. Test the seeds of castor oil, or the leaves of mint for oils, carrying out the grease spot test on a piece of filter paper. (See 20.1.2.)

(c) *Calcium oxalate*, Examine transverse sections of leaves of rhubarb, horse chestnut or other deciduous tree leaves obtained in autumn. Crystals, called raphides, of calcium oxalate are often present in the cells.

23 Experiments on Growth and Reproduction

Living organisms acquire material from their environment, and, with the energy they obtain from respiration, they build up their bodies, and ultimately reproduce.

23.1 GROWTH

ANIMALS

i. Flatworms (*Planaria* spp.). See 7.1

Keep specimens supplied with food, in dishes at different temperatures. Measure their lengths over a period of time, and record your results.

ii. Cabbage White Butterfly Caterpillars (*Pieris brassicae*). See 10.2.2

Keep specimens on cabbage leaves, and record size after each moult.

iii. Trout (*Salmo fario*)

Record the growth in length of an alevin immediately after hatching for a period of several weeks.

iv. Frog (*Rana temporaria*). See 13.2

Keep tadpoles in dishes at different temperatures. Measure and record the increase in size over a period of time.

v. Man

Make a small scratch near the base of a finger nail, and measure the rate of growth of the mark away from the base of the nail at weekly intervals.

vi. Mouse

Weigh a newly born mouse, and keep a record of the increase in weight over a period of time.

vii. The influence of hormones on growth

Set up four experiments, by placing 10 tadpoles in each of four jars of water. In jar 1, place water weeds only. Jar 2, water weeds, and half a thyroid tablet added twice per week. Jar 3, a small portion of fresh meat added twice per week. Jar 4, a small portion of fresh meat and half a thyroid tablet added twice per week. Keep the water clean in all the jars, and remove any dead tadpoles. A little nipagen added to the water will help to prevent the growth of fungi and bacteria. Observe and record the rate of growth of the tadpoles in each of the four experiments.

PLANTS

i. Growth of a whole plant

(*a*) Observe the rate of growth of a plant shoot such as a potted geranium, by attaching the growing tip to a growth lever by means of a light cord. Record the results by means of a graph, plotting increase in length of the plant shoot against time. Observe any difference between growth by day and night.

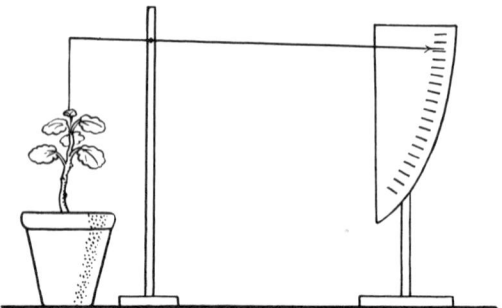

(*b*) Study the increase in dry weight over a period of time. Weigh separately seeds of pea, bean, maize and sunflower. Soak them in water, germinate and pot out the seedlings. When the resulting plants have matured, collect each one to include all its roots, leaves and stem. Dry each plant in an oven at about 101°C, and then weigh each specimen. Record the increase in dry weight as compared with the dry weight of the original seed.

ii. Growth in roots

Observe the rate of growth of pea roots by growing peas in gas jars between the glass and a piece of blotting paper held in place by damp moss inside the jar. Place experiments in different temperatures, and compare the rates of growth.

iii. Region of growth

Observe the region of growth in the roots of peas or broad beans. Germinate the seeds, and when the radicle is 2 cm long dry it, and make marks 1 mm apart by laying a thread soaked in Indian ink across it at 1 mm intervals. When the ink is dry, set the seedlings in damp sawdust. After 24 hours record the distance between each mm mark, and so determine the region of growth.

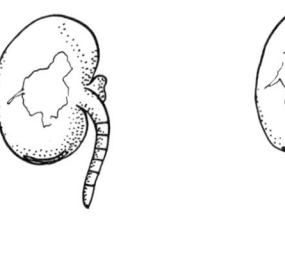

GROWTH IN 24 HOURS

iv. Growth in leaves

Select a young leaf, such as sycamore, in its natural position on the tree. Lay the leaf against a piece of graph paper and draw its outline. Mark the leaf with a piece of coloured wool, so that you can easily identify it again. Record its outline on the same piece of graph paper at successive intervals.

v. Growth hormones

(*a*) Obtain a selective weed killer such as 2-4-D, and make a solution of the strength recommended on the bottle. Mark out two squares of metre side on a lawn. Water one square regularly with pure water. Water the other at the same time with the 2-4-D solution. Record the results obtained.

(*b*) Grow broad-leaved plants, such as buttercups, in plant pots. Spray each pot with 2-4-D solutions of varying strengths, and observe the effects after a few days. Examine the cells of a leaf removed from each plant at different stages in the experiment.

(*c*) Take two cuttings of a geranium plant. Place one in damp sawdust, and another in damp sawdust after first treating the roots with a rooting hormone. After one week, carefully remove the plants, and compare the control with the one treated with hormone. (See page 79.)

(*d*) To 1000 cm³ of distilled water add 20 mg sucrose, a mineral nutrient tablet and 10·0 mg of indole-3-acetic acid (I.A.A.). Place 250 cm³ of this solution in a beaker, and use the remainder to produce 3 portions, successively diluted 10 times, to give the following strengths

·01 mg/1000 cm³
0·1 mg/1000 cm³
1·0 mg/1000 cm³
10·0 mg/1000 cm³ (i.e. the original solution)

Place some of each one in a petri dish over a piece of graph paper. Into each of the prepared solutions put the cut end of a maize coleoptile about 2 cm long, and measure its length by means of graph paper. Record any changes in length which occur in each specimen over a period of time.

(e) Grow some maize seedlings until the coleoptiles are about 2 cm long and divide them into four groups as follows—

1. Control.
2. Cut off the tips (the terminal 2 mm).
3. Cut off the tips and after about 5 minutes replace one on the stump of each coleoptile.
4. Cut off the tips and stand them with their bases on plain agar jelly for about 5 hours. Then place about 2 mm³ of agar from under each tip back on each stump. Record the behaviour of each of the four groups of seedlings.

vi. Growth in Fungi

(a) **Yeast** (*Saccharomyces* **spp.**) Place approximately 15 cm of 20% glucose solution in a test tube to which about 5 small grains of yeast have been added. Put it in a beaker of water to act as a water bath at 25°C. Stir in the yeast powder thoroughly with a glass rod, and then withdraw one drop on to a microscope slide and cover with a coverslip. Try to estimate the number of cells visible in the high power field of view. If a blood count apparatus is available (See 21.1.ix), the accuracy of the observations will be much improved by counting the number of yeast cells visible in a definite number of squares.

Leave the yeast suspension for 30 minutes keeping it in the water bath at 25°C and then stir it with the glass rod, and again withdraw one drop of solution with a pipette. Count the number of yeast cells and determine whether there has been any change in the density of the population. Carry out similar experiments at temperatures of 5°C, 15°C and 35°C to determine whether temperature has any influence on the rate of growth.

(b) **Pin Mould** (*Rhizopus* **or** *Mucor* **spp**) Introduce a small piece of mould into the centre of a petri dish containing nutrient agar and incubate it at a steady temperature. Measure the rate of increase in the diameter of the mould growth over a period of time.

23.2 REGENERATION

ANIMALS

Many animals have the ability to regenerate parts of their bodies, if these have been lost or damaged. Animals in the simpler Phyla have greater powers of regeneration than those in more complex forms. The latter have, however, the power to heal wounds.

i. Flatworms (*Planaria* spp.). See 7.1

Select a number of specimens, place them one at a time on damp filter paper, and cut each one with a sharp scalpel, varying the position and direction of the cut. Return the animals to dishes of water, and record their ability to regenerate new body tissue over a period of days.

ii. Earthworm (*Lumbricus* spp.). See 9.1

Take a number of specimens, and cut each one transversely in different positions. Place each one in a small wormery and examine it from time to time, and record the powers of regeneration of the different specimens.

iii. Crustacea. See 10.1

Many crustaceans have the power to regenerate whole limbs. Observe various specimens of crabs, shrimps, etc., for evidence of regenerated limbs.

iv. Starfish (*Asterias* spp.). See 12.1

Specimens are frequently found with one or more arms smaller than the rest. These are in the course of regenerating damaged parts.

v. Lizard (*Lacerta* spp.). See 13.3

These animals have a breaking plane through one of the vertebrae in the tail, and the tail may break off if it is seized by a bird. A new, rather stunted tail frequently grows in its place. Look out for specimens showing this phenomenon.

PLANTS

Many plants have high powers of regeneration. When a plant is wounded, the exposed surface respires more rapidly and new tissues are quickly formed.

i. Wounding plant tissues

(a) Grow specimens of grass in a seed-box, and note the effect of frequent cutting of the leaves with a pair of scissors. (Tillering.)

(b) Observe specimens of dandelions. Cut a plant off close to the ground, and observe its powers of regeneration.

(c) Examine oak apples. These are caused by the wound due to the entry of a small insect into the tissues of the tree.

(d) Galls. Examine galls on the leaves of various trees, which are growths caused by the entry of insects into the tissues.

ii. Grafting

The success of this process depends on the ability of the cut portion of a plant shoot to regenerate. Twigs of apple or pear are suitable subjects for grafting experiments.

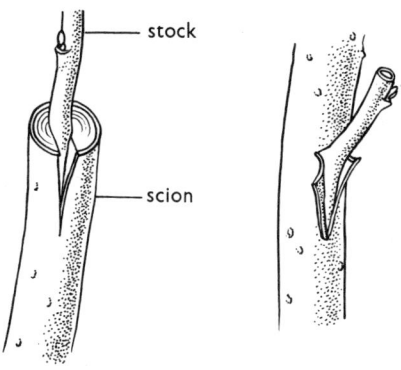

TYPES OF GRAFTING

iii. Cuttings

Take cuttings of a geranium, by slicing down the surface of a stem near its junction with a petiole, and remove the leaf, so that the periole has a portion of stem still attached to it. Plant such cuttings in pots of well-watered soil. Dip each cutting in rooting hormone to hasten the growth of new roots from the cut surface.

23.3 REPRODUCTION

23.3.1 Asexual

ANIMALS

Examine prepared slides of *Amoeba* (see 5.1) and *Paramecium* (see 5.2).

PLANTS

(i) Examine specimens of *Pleurococcus*. (See Chlorophyta 14.1.1.)
(ii) Examine the sporanglophores from a culture of *Mucor*. (See Phycomycota 14.4.1.)
(iii) Examine the zoosporangia of *Saprolegnia*. (See 14.4.1.)
(iv) Examine the spore capsules of *Funaria*. (See Bryophyta 15.1.)
(v) Examine the sporangia of the male fern. (See Pterophyta 16.1.)

23.3.2 Vegetative

This process involves the detachment of some part of the body of an organism, which then develops into a new complete specimen.

ANIMALS

(i) Examine specimens of *Hydra* spp. for budding. (See 6.1.)

PLANTS

(i) Examine specimens of *Spirogyra* spp. (See Chlorophyta 14.1.3.)
(ii) Examine yeast cells under the microscope, and observe the formation of buds. (See Ascomycota 14.5.1.)

(iii) Observe the methods employed by flowering plants (Anthophyta), as in creepers, runners, suckers, storage organs. (See pages 79, 80 and 81.)

(iv) Grow specimens of Duckweed (*Lemna minor*) in a jam jar of water with a small amount of soil at the bottom. Observe their vegetative method of growth over a period of several days.

23.3.3 Sexual

This process involves the fusion of two gametes to produce a zygote, which then develops into the adult stage of the organism.

ANIMALS

i. *Paramecium* spp. See 5.2

Examine prepared microscope slides showing conjugation.

ii. *Hydra* spp. See 6.1

Examine specimens of *Hydra* for the presence of testes. Sperms can often be seen moving inside a mature testis.

iii. Earthworm (*Lumbricus* spp.). See 9.1

Dissect a freshly killed Earthworm, remove a seminal vesicle, and make a smear of the sperms on a microscope slide. Examine this under the low and high powers of the microscope.

iv. Marine Worm (*Pomatoceros triqueter*). See 9.4

Obtain some specimens of this animal, cut off the terminal 2 mm at the pointed end of its tube and push a seeker into the blunt end of the tube, and so force the worm out through the sharp end. The sexes are separate the males being pale yellow and the females red or violet in colour. Place a male and a female worm together in a petri dish of sea water as soon as they have been removed from their tubes, and within a few minutes the sea water should contain their sperms and eggs. Pipette a drop of the sea water into a cavity slide, and observe fertilization and the early development of the zygotes under the microscope.

v. Locust (*Locusta migratoria*). See 10.2.6

Dissect out the testis and the fat which surrounds it from the abdomen of a male Locust. Place the material in a drop of saline on a slide and gently tease out the testis. Place a few testis follicles in saline on a fresh slide, and gently squash them with a scalpel blade. Remove excess saline with a piece of blotting paper and then add a drop of orcein stain. Place a coverslip over the material and press it gently at the same time removing excess liquid with a piece of blotting paper. Warm the slide for a few seconds and examine it under the high power of the microscope when it has cooled. Examine the testis cells which are visible in various stages of development.

vi. Starfish (*Asterias* spp.). See 12.1

If living specimens can be obtained, it is comparatively easy to remove the gametes and watch fertilization under the microscope. Dissect out the testes from a male starfish and place them in a dish of sea water. Similarly remove the ovaries from a female. Place a few eggs in a cavity slide in a drop of sea water. With a pipette pick up some of the sea water from the dish containing the testes. Squirt a few drops of this water containing the sperms over the eggs on the cavity slide. Observe at frequent intervals, when sperm may be seen surrounding the eggs. After fertilization has occurred look for stages in the development of larvae.

vii. Herring (*Clupea harengus*). See 13.1.1

Examine the soft roe (testis), and the hard roe (ovary) by making squash preparations on a microscope slide.

viii. Rat (*Rattus norvegicus*). See page 55.

Examine prepared slides of testis and ovary.

PLANTS

i. Fusion of isogametes

This takes place when gametes of the same size fuse together to produce a zygote.

Observe prepared slides showing conjugation in *Spirogyra* (see page 57), and *Mucor* (see page 60).

ii. Fusion of anisogametes

This takes place when two gametes, each of different size, fuse together to form a zygote.

(*a*) Seaweeds (*Fucus* spp.). See 14.2. Examine fresh specimens of *Fucus* spp., cut sections through the conceptacles and examine antheridia and oogonia. Wipe the slime from the outside of the conceptacles in mature specimens of *Fucus*, and mount on a slide. Examine under the high power of the microscope for the presence of gametes.

(*b*) Male Fern (*Dryopteris* spp.). See 16.1. Mount a fern prothallus in a drop of water on a slide, and cover with a coverslip. Examine under the high power; male gametes can often be seen swimming in the water.

(*c*) Flowering plants (Anthophyta). Collect pollen from a variety of flowers. Mount them in drops of sugar solution (strengths 2%–10%) on microscope slides. Leave these under a bell jar for a few hours, and examine them for the formation of pollen tubes.

23.4 DEVELOPMENT

Observe the manner in which various organisms develop from the fertilized zygote into the adult condition.

ANIMALS

i. Earthworm (*Lumbricus* spp.). See 9.1

Carefully pass specimens of fresh soil through a wire-mesh sieve, and examine it for the presence of worm cocoons, which are about the size of barley grains. Collect these, and place them in a small wormery. Observe their subsequent development.

ii. Pond Snail (*Limnaea* spp.). See 11.2

Collect the eggs of a pond snail and mount them on a cavity slide. Observe under the low power of the microscope the various stages of the development of the young snail within the egg shell.

iii. Trout (*Salmo* spp.)

Obtain trout eggs from a hatchery early in January. Place them in a tank with gravel at the bottom, and pass a steady current of water through it. Observe the various stages of development, the rate depending on the temperature of the water. Shortly after hatching, examine specimens under the low power of the microscope in a watch-glass. Circulation of the blood can be observed in the gills.

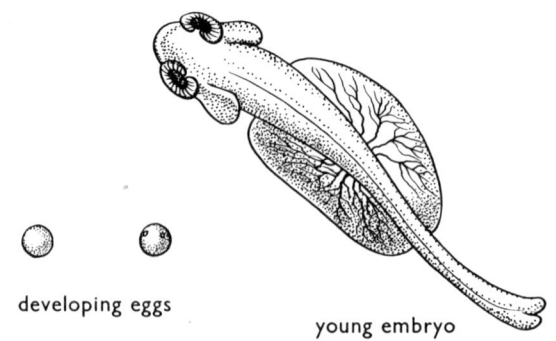

developing eggs

young embryo

DEVELOPMENT OF TROUT

iv. Frog (*Rana temporaria*). See 13.2.1

Keep specimens of frog spawn in small tanks. Examine the various stages of development at intervals of a few days.

v. Hen's eggs

Obtain fertile eggs from a farmer, and with a pencil mark a cross on one side of the shell. Place the eggs in an incubator (a laboratory electric oven is satisfactory) and incubate at a steady temperature of 39°C. Place a small jar of water in the incubator to maintain the humidity of the atmosphere. The eggs should be turned through 180° twice per day in order to prevent the embryo sticking to the shell. The pencil cross marks on the shells indicate the position to place each egg, after it has been turned. The developing embryo is most likely to stick to the inside of the shell on about the seventh day, so it is important to ensure that the eggs are turned regularly at about this stage of incubation. Open eggs at successive intervals, of about 36 hr., 48 hr, 72 hr, 4, 7, 10 and 15 days of incubation.

To dissect, remove an egg from the incubator and immerse it in a large pie-dish of water at about 40°C. It is helpful to make a plaster of Paris mould the shape of the egg, in which it can be rested in the pie-dish during the dissection. With a pair of sharp-pointed scissors, snip a window about 3 cm square in the side of the shell, and remove the shell and its membranes. It is best to have one person holding the egg in position while the other carries out the dissection.

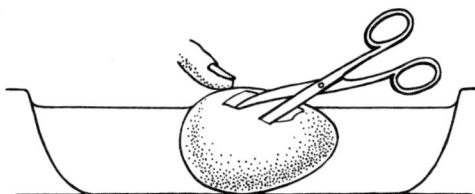

At the 36, 48 and 72 hr stages, observe the developing embryo in position with a hand lens. Subsequently dissect out the embryo, and mount it in a drop of water on a cavity slide, to examine it under the low power of the microscope. Later stages will be too large to examine under the microscope.

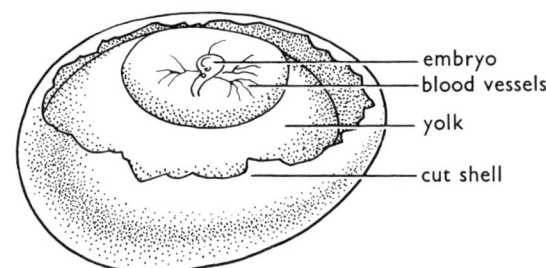

HEN'S EGG. 3 DAYS INCUBATION

vi. Rat (*Rattus* spp.). See 13.5.1

Examine embryos removed from the uterus of a pregnant female, using a hand lens.

PLANTS

Study the development of flowing plants (Anthophyta) by observing the manner in which seeds germinate.

Seeds are best germinated in small gas jars lined with a cylinder of damp blotting paper pressed against the sides with damp moss. Plant a wide variety of seeds of different plants, and observe their methods of germination.

Carry out experiments to determine the various factors in the environment which influence the manner in which seeds germinate.

i. To show that water is necessary

Grow four groups of pea seeds in each of the following conditions:
(a) Dry seeds on dry cotton-wool.
(b) Seeds soaked for 24 hr on dry cotton-wool.
(c) Dry seeds on damp cotton-wool.
(d) Seeds soaked for 24 hr on damp cotton-wool.

Place all the seeds in dishes at an even temperature, and record the results.

ii. To show that oxygen is necessary

Place some pea seeds on damp cotton-wool, soaked in water which has previously been boiled to remove the dissolved gases. Place them on a glass plate under a bell jar also covering a candle. Light this, before firmly fixing the bell jar to the glass plate with vaseline. The candle will go out when all the oxygen has been consumed, so leaving the seeds in an oxygen free atmosphere. Set up a control experiment, from which oxygen has not been removed.

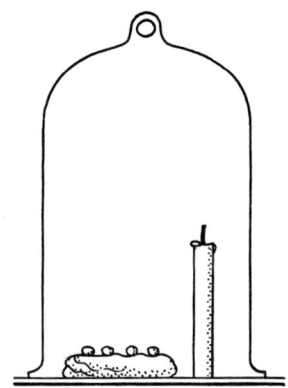

iii. The effect of temperature

Soak some pea seeds for 24 hr. Germinate them in gas gars at different temperatures, and record the rate of growth of the roots in each temperature over a few days.

iv. The effect of light

Germinate some peas previously soaked for 24 hr, one set under normal conditions, and the other in darkness. Observe the two sets over a period of time, and compare their manner of development.

v. Percentage germination

Place about 50 peas in a large dish of damp sawdust and count the number which germinate at different intervals of time. Record your results as a percentage of the total number. Repeat the experiments with other types of seeds.

vi. Shepherd's Purse (*Capsella bursa-pastoris*)

Remove some seeds from the young fruit (silicula) with a pair of forceps, and lay them on a microscope slide. Pinch them gently with the forceps, when the embryo inside the seed will be expelled. Examine an embryo in a drop of water under the low power of the microscope. Also examine a prepared slide.

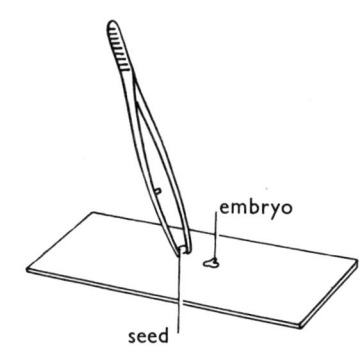

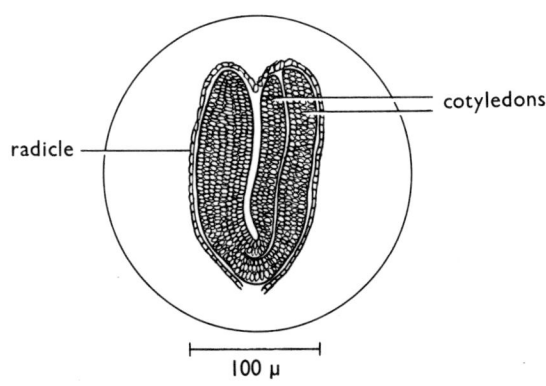

EMBRYO MAGNIFIED

119

23.5 HEREDITY

i. Experiments on heredity in the fruit fly (*Drosophila melanogaster*)

The fruit fly is an ideal organism in which to study the principles of genetics. It is a) easy to handle, b) easy to determine the sex of the adult flies. The tip of the male abdomen appears as a black blob, whereas the female abdomen is more pointed and has distinct black stripes. c) Kept in an incubator at 25°C the life cycle is completed in about 14 days. d) A few hours after females emerge from the pupa cases they are ready to be mated. e) The larvae or grubs can be fed on a culture medium of agar jelly containing banana, yeast and a fungicide. This medium can be obtained ready made up from laboratory suppliers.

FEMALE FLY   MALE FLY

Crossing long and vestigeal winged types. It is essential that all apparatus used should be thoroughly sterilized before use in a pressure cooker or autoclave. Obtain pure strains of flies and the culture medium from laboratory suppliers. Prepare the food medium and pour it while hot into sterile specimen tubes, and allow them to cool. If the medium is not supplied with fungicide and yeast, put a little of each into the jelly before it sets. Place a small sliver of paper handkerchief vertically in the medium, for the larvae to crawl up just before pupation.

Using an etherizer (as described below) select one male and one female fly and place them in a specimen tube prepared as described and cork it up with a porous bung (or sterilized cotton wool). After mating the female will lay a hundred or more eggs on the surface of the jelly. When larvae emerge remove the parent flies and destroy them to pre-

vent possible interference with other experiments should they escape into the laboratory. Watch the course of the life cycle; there are 3 larval stages before the pupa is formed. When the adult flies of the F_1 generation emerge, they must be investigated as soon as possible, for the females become fertile within a few hours.

Technique for handling and counting flies

1. To anaesthetize the flies, empty them into an etherizer. This can be made from a specimen tube, cork and funnel, together with a small amount of cotton wool soaked in ether. (See diagram). A very small amount of ether should be placed on the cotton wool from a pipette, an excess of the fluid is likely to kill the flies.

2. Shake the anaesthetized flies from the etherizer into a petri dish over a white background, and using a camel hair brush separate them into types, e.g. male, female, long and vestigeal wings. If the flies start to come round, cover the petri dish with its lid to which a small piece of cotton wool soaked in ether has been attached.

3. Set up further experiments to investigate the results of mating F_1 male and female flies, and so obtain F_2 offspring.

ii. Dominance in human beings

(*a*) Tasting phenylthiourea (P.T.C.). Some people are able to taste a substance called phenylthiourea and recognize it as bitter, others do not taste it at all. This seems to be due to the interaction of a simple dominant and recessive gene mechanism.

Prepare a solution of P.T.C. containing 1·3 g/1000 cm³ boiled tap water which has been allowed to cool. Soak a small piece of filter paper in the solution and place it on the tongue of a subject for a few seconds. Follow this with a similar test using a piece of filter paper soaked in boiled tap water only, which has been allowed to cool. Determine how many people in a sample can distinguish between the P.T.C. solution and the control. Record and analyse your results. It is advisable for each subject to wash the mouth out with water, before and after the test.

(*b*) Tongue rolling. Some people can place the tongue between the lips and roll it, some cannot. The former characteristic seems to be dominant to the latter. Test this phenomenon in a sample of people.

(*c*) Record the variety of finger prints in a number of individuals. Smooth a few drops of finger print ink out on to a piece of polished metal plate. Then lightly roll a clean finger on the ink from one side to the other, and transfer the impression on to a piece of clean white paper, making a similar rolling movement.

(*d*) Investigate the distribution of ABO blood groups in a number of individuals, using the technique described in 21.1.ix(f).

iii. Laws of chance in heredity

To illustrate the laws of chance operating in simple mendelian experiments, carry out a class experiment in which each person spins two pennies, and records how they fall, for a number of experiments. Count the number of times the pennies fall in the following combinations:

> Head, head.   (Pure dominant.)
> Head, tail.   (Hybrid.)
> Tail, tail.   (Pure recessive.)

Add together the results produced from each member of the class. It is best to have at least a thousand results, from the combined experiments in the class.

23.6 VARIATION

i. Variation in human beings

(*a*) *Cephalic index.* Record this in a number of individuals, by measuring the head with a pair of large calipers. Take one measurement from the mid-line of the forehead to the projection in the mid-line at the back of the head. Take the other at right angles at the widest point of the head, usually just above the ears. Record the cephalic index as:

$$\text{Cephalic index} = \frac{\text{Width} \times 100}{\text{Length}}$$

Below 75—the skull is long headed, 75–80—round headed, above 80—broad headed.

(*b*) Eyes. Observe, and record the colour of the iris in a number of individuals.

(*c*) *Nasal index.* Measure the length of the nose from the mid-line at the base of the nasal septum to the deepest indentation between the eyes. Measure the width at the base of the nose. Record the index in a number of individuals.

$$\text{Nasal index} = \frac{\text{Width} \times 100}{\text{Length}}$$

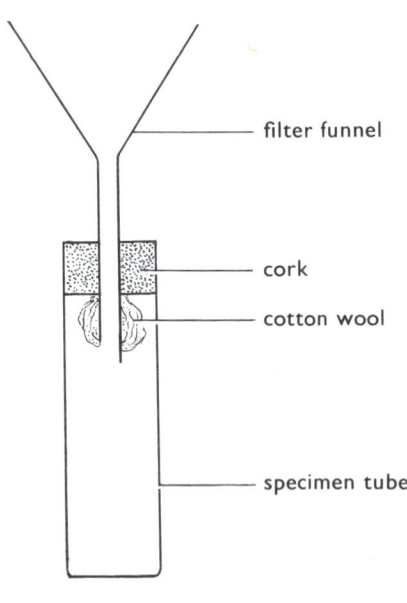

filter funnel

cork

cotton wool

specimen tube

ETHERIZER

ii. Variation in maize fruits

Make a record of the main types of fruits recognizable in a single maize cob.

iii. Variation in broad beans

Record the variations in lengths of broad beans, measuring the total length of each seed with a pair of dividers. Record the number of beans in each of a number of size groups.

iv. Variation in composite flowers

Record the number of florets present in the heads of about a dozen dandelion flowers.

v. Chromosomes

These may be studied by the technique of squashing tissue under a coverslip on a microscope slide as described for Locust testis. (See 23.3.3.)

Cut off about 5 mm from the tip of a growing root of a young onion plant and place it in a watch glass containing 1 : 10 mixture of N.Hydrochloric acid and acetic orcein stain. Warm for about 5 minutes, then transfer the root tip to a slide and add a few drops of acetic orcein stain.

Gently squash the tissue under a coverslip on a slide and warm it for a few seconds. When cool examine the preparation under the high power of the microscope.

vi. Deoxyribose nucleic acid (DNA)

Make a simple model of the structure of part of a DNA molecule to show how the nucleotides are assembled. A two dimensional model can be made from cards cut to shapes to represent the main molecules. Phosphate can be represented by a circle, ribose by a pentagon. Of the bases, cytosine and guanine pair together and a suitable interlocking shape should be cut to represent this. Adenine and thymine similarly always pair together, and this should be represented by a different outline for the interlocking.

A three dimensional model can be made from a piece of thick dowelling fixed in a small wooden base. Holes should be drilled in the dowelling to take small pieces of cylindrical wood, each arranged at an angle to represent the helical shape of the DNA molecule. Stick two strands of tape to represent the phosphate bonding and fix the pentagon shapes (representing ribose) to the wooden frame members with drawing pins. Stick base pairs cytosine : guanine and adenine : thymine across the wooden frame members with adhesive.

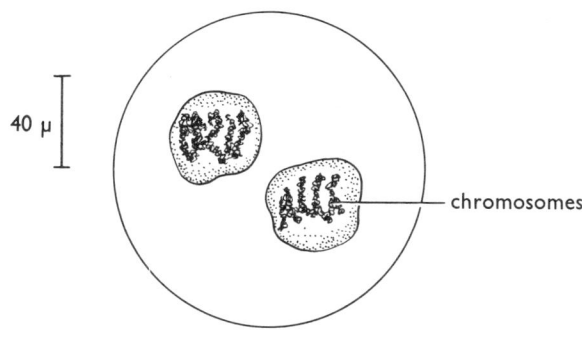

CELLS FROM YOUNG ROOT TIP OF ONION

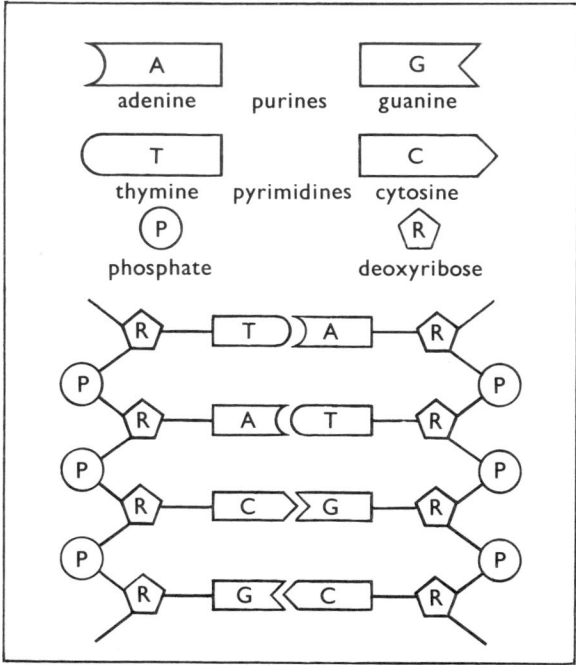

DNA   TWO DIMENSIONAL MODEL

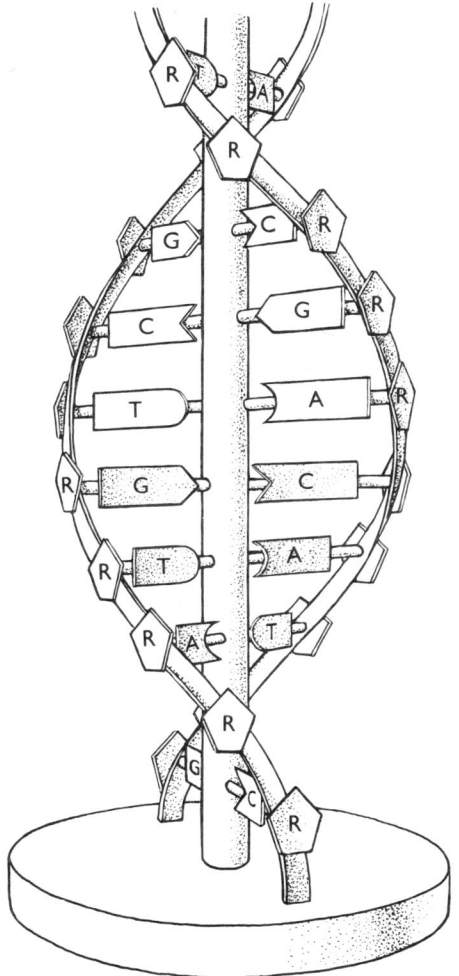

DNA   THREE DIMENSIONAL MODEL

Index